Calculus Workbook

FOR

DUMMIES®

by Mark Ryan

Other *For Dummies* math titles:

Algebra For Dummies 0-7645-5325-9
Algebra Workbook For Dummies 0-7645-8467-7
Calculus For Dummies 0-7645-2498-4
Geometry For Dummies 0-7645-5324-0
Statistics For Dummies 0-7645-5423-9
Statistics Workbook For Dummies 0-7645-8466-9
TI-89 Graphing Calculator For Dummies 0-7645-8912-1 (also available for TI-83 and TI-84 models)
Trigonometry For Dummies 0-7645-6903-1
Trigonometry Workbook For Dummies 0-7645-8781-1

WILEY

Wiley Publishing, Inc.

Calculus Workbook For Dummies®

Published by
Wiley Publishing, Inc.
111 River St.
Hoboken, NJ 07030-5774
www.wiley.com

Copyright © 2005 by Wiley Publishing, Inc., Indianapolis, Indiana

Published by Wiley Publishing, Inc., Indianapolis, Indiana

Published simultaneously in Canada

For general information on our other products and services, please contact our Customer Care Department within the U.S. at 877-762-2974, outside the U.S. at 317-572-3993, or fax 317-572-4002.

For technical support, please visit www.wiley.com/techsupport.

Wiley also publishes its books in a variety of electronic formats. Some content that appears in print may not be available in electronic books.

Library of Congress Control Number is available from the publisher.

ISBN-13: 978-0-7645-8782-5

ISBN-10: 0-7645-8782-X

Manufactured in the United States of America

10 9

1B/RV/QY/QV/IN

WILEY

About the Author

A graduate of Brown University and the University of Wisconsin Law School, **Mark Ryan** has been teaching math since 1989. He runs the Math Center in Winnetka, Illinois (www.the mathcenter.com), where he teaches high school math courses including an introduction to calculus and a workshop for parents based on a program he developed, *The 10 Habits of Highly Successful Math Students*. In high school, he twice scored a perfect 800 on the math portion of the SAT, and he not only knows mathematics, he has a gift for explaining it in plain English. He practiced law for four years before deciding he should do something he enjoys and use his natural talent for mathematics. Ryan is a member of the Authors Guild and the National Council of Teachers of Mathematics.

Calculus Workbook For Dummies is Ryan's third book. *Everyday Math for Everyday Life* was published in 2002 and *Calculus For Dummies* (Wiley) in 2003.

A tournament backgammon player and a skier and tennis player, Ryan lives in Chicago.

Author's Acknowledgments

My agent Sheree Bykofsky of Sheree Bykofsky Associates, Inc., has represented me now on three successful books. Her ability to make connections in the publishing world and to close deals — and to see the forest when I might be looking at the trees — has made her invaluable.

Many thanks to my bright, professional, and computer-savvy assistants, Benjamin Mumford, Randy Claussen, and Caroline DeVane. Josh Dillon did a meticulous review of the mathematics in the book. He knows calculus and how to communicate it clearly. And a special thanks to the multi-talented Amanda Wasielewski who did everything from typing the book's technical equations to creating diagrams to keeping the project organized to checking the calculus content.

Gene Schwartz, President of Consortium House, publishing consultants, and an expert in all aspects of publishing, gave me valuable advice for my contract negotiations. My friend, Beverly Wright, psychoacoustician and writer extraordinaire, was also a big help with the contract negotiations. She gave me astute and much needed advice and was generous with her time. I'm grateful to my consultant, Josh Lowitz, Adjunct Associate Professor of Entrepreneurship at the University of Chicago School of Business. He advised me on every phase of the book's production. His insights into my writing career and other aspects of my business, and his accessibility, might make you think I was his only client instead of one of a couple dozen.

This book is a testament to the high standards of everyone at Wiley Publishing. Special thanks to Joyce Pepple, Acquisitions Director, who handled our contract negotiations with intelligence, honesty, and fairness, and to Acquisitions Editor Kathy Cox, who deftly combined praise with a touch of gentle prodding to keep me on schedule. Technical Editor Dale Johnson did an excellent and thorough job spotting and correcting the errors that appeared in the book's first draft — some of which would be very hard or impossible to find without an expert's knowledge of calculus. The layout and graphics team did a fantastic job with the book's thousands of complex equations and mathematical figures.

Finally, the book would not be what it is without the contributions of Project Editor Laura Peterson-Nussbaum. She's an intelligent and skilled editor with a great sense of language and balance. And she has the ability to tactfully suggest creative changes — most of which I adopted — without interfering with my idiosyncratic style. It was a pleasure to work with her.

Publisher's Acknowledgments

We're proud of this book; please send us your comments through our Dummies online registration form located at www.dummies.com/register/.

Some of the people who helped bring this book to market include the following:

Acquisitions, Editorial, and Media Development

Project Editors: Laura Peterson-Nussbaum

Acquisitions Editor: Kathy Cox

Editorial Program Assistant: Courtney Allen

Technical Editor: Dale Johnson

Senior Permissions Editor: Carmen Krikorian

Editorial Manager: Michelle Hacker

Editorial Supervisor: Carmen Krikorian

Editorial Assistants: Hanna Scott, Melissa Bennett

Cartoons: Rich Tennant (www.the5thwave.com)

Composition Services

Project Coordinator: Adrienne Martinez

Layout and Graphics: Jonelle Burns, Andrea Dahl, Kelly Emkow, Carrie A. Foster, Denny Hager

Proofreaders: Laura Albert, Jessica Kramer, Joe Niesen, Christine Pingleton

Indexer: Richard T. Evans

Publishing and Editorial for Consumer Dummies

    Diane Graves Steele, Vice President and Publisher, Consumer Dummies

    Joyce Pepple, Acquisitions Director, Consumer Dummies

    Kristin A. Cocks, Product Development Director, Consumer Dummies

    Michael Spring, Vice President and Publisher, Travel

    Kelly Regan, Editorial Director, Travel

Publishing for Technology Dummies

    Andy Cummings, Vice President and Publisher, Dummies Technology/General User

Composition Services

    Gerry Fahey, Vice President of Production Services

    Debbie Stailey, Director of Composition Services

Contents at a Glance

Introduction 1

Part I: Pre-Calculus Review 5

Chapter 1: Getting Down the Basics: Algebra and Geometry 7
Chapter 2: Funky Functions and Tricky Trig 19

Part II: Limits and Continuity 29

Chapter 3: A Graph Is Worth a Thousand Words: Limits and Continuity 31
Chapter 4: Nitty-Gritty Limit Problems 39

Part III: Differentiation 57

Chapter 5: Getting the Big Picture: Differentiation Basics 59
Chapter 6: Rules, Rules, Rules: The Differentiation Handbook 69
Chapter 7: Analyzing Those Shapely Curves with the Derivative 91
Chapter 8: Using Differentiation to Solve Practical Problems 123

Part IV: Integration and Infinite Series 157

Chapter 9: Getting into Integration 159
Chapter 10: Integration: Reverse Differentiation 177
Chapter 11: Integration Rules for Calculus Connoisseurs 193
Chapter 12: Who Needs Freud? Using the Integral to Solve Your Problems......219
Chapter 13: Infinite Series: Welcome to the Outer Limits 243

Part V: The Part of Tens 263

Chapter 14: Ten Things about Limits, Continuity, and Infinite Series 265
Chapter 15: Ten Things You Better Remember about Differentiation 269
Chapter 16: Ten Things to Remember about Integration If You Know What's Good for You 273

Index 277

Table of Contents

Introduction 1

About This Book 1
Conventions Used in This Book 1
How to Use This Book 2
Foolish Assumptions 2
How This Book Is Organized 2
    Part I: Pre-Calculus Review 2
    Part II: Limits and Continuity 3
    Part III: Differentiation 3
    Part IV: Integration and Infinite Series 3
    Part V: The Part of Tens 3
Icons Used in This Book 4
Where to Go from Here 4

Part I: Pre-Calculus Review 5

Chapter 1: Getting Down the Basics: Algebra and Geometry 7

Fraction Frustration 7
Misc. Algebra: You Know, Like Miss South Carolina 9
Geometry: When Am I Ever Going to Need It? 12
Solutions for This Easy Elementary Stuff 15

Chapter 2: Funky Functions and Tricky Trig 19

Figuring Out Your Functions 19
Trigonometric Calisthenics 22
Solutions to Functions and Trigonometry 25

Part II: Limits and Continuity 29

Chapter 3: A Graph Is Worth a Thousand Words: Limits and Continuity ... 31

Digesting the Definitions: Limit and Continuity 31
Taking a Closer Look: Limit and Continuity Graphs 34
Solutions for Limits and Continuity 37

Chapter 4: Nitty-Gritty Limit Problems 39

Solving Limits with Algebra 39
Pulling Out Your Calculator: Useful "Cheating" 44
Making Yourself a Limit Sandwich 46
Into the Great Beyond: Limits at Infinity 47
Solutions for Problems with Limits 50

Part III: Differentiation 57

Chapter 5: Getting the Big Picture: Differentiation Basics 59

The Derivative: A Fancy Calculus Word for Slope and Rate 59
The Handy-Dandy Difference Quotient 61
Solutions for Differentiation Basics 64

Chapter 6: Rules, Rules, Rules: The Differentiation Handbook **69**

Rules for Beginners..69
Giving It Up for the Product and Quotient Rules 72
Linking Up with the Chain Rule 75
What to Do with Ys: Implicit Differentiation..................................78
Getting High on Calculus: Higher Order Derivatives 80
Solutions for Differentiation Problems 82

Chapter 7: Analyzing Those Shapely Curves with the Derivative **91**

The First Derivative Test and Local Extrema 91
The Second Derivative Test and Local Extrema 95
Finding Mount Everest: Absolute Extrema 98
Smiles and Frowns: Concavity and Inflection Points.....................102
The Mean Value Theorem: Go Ahead, Make My Day 106
Solutions for Derivatives and Shapes of Curves 108

Chapter 8: Using Differentiation to Solve Practical Problems....................**123**

Optimization Problems: From Soup to Nuts 123
Problematic Relationships: Related Rates......................................127
A Day at the Races: Position, Velocity, and Acceleration 131
Make Sure You Know Your Lines: Tangents and Normals..................134
Looking Smart with Linear Approximation.....................................138
Solutions to Differentiation Problem Solving 140

Part IV: Integration and Infinite Series *157*

Chapter 9: Getting into Integration **159**

Adding Up the Area of Rectangles: Kid Stuff 159
Sigma Notation and Reimann Sums: Geek Stuff 162
Close Isn't Good Enough: The Definite Integral and Exact Area 166
Finding Area with the Trapezoid Rule and Simpson's Rule 168
Solutions to Getting into Integration 171

Chapter 10: Integration: Reverse Differentiation **177**

The Absolutely Atrocious and Annoying Area Function....................177
Sound the Trumpets: The Fundamental Theorem of Calculus 179
Finding Antiderivatives: The Guess and Check Method 183
The Substitution Method: Pulling the Switcheroo............................185
Solutions to Reverse Differentiation Problems 188

Chapter 11: Integration Rules for Calculus Connoisseurs **193**

Integration by Parts: Here's How u du It.......................................193
Transfiguring Trigonometric Integrals 196
Trigonometric Substitution: It's Your Lucky Day! 198
Partaking of Partial Fractions..201
Solutions for Integration Rules..205

Chapter 12: Who Needs Freud? Using the Integral to Solve Your Problems **219**

Finding a Function's Average Value 219
Finding the Area between Curves 220
Volumes of Weird Solids: No, You're Never Going to Need This 222
Arc Length and Surfaces of Revolution...227
Getting Your Hopes Up with L'Hôpital's Rule 229

Disciplining Those Improper Integrals...231
Solutions to Integration Application Problems..234

Chapter 13: Infinite Series: Welcome to the Outer Limits **243**

The Nifty nth Term Test 243
Testing Three Basic Series 245
Apples and Oranges . . . and Guavas: Three Comparison Tests 247
Ratiocinating the Two "R" Tests...251
He Loves Me, He Loves Me Not: Alternating Series.....................................253
Solutions to Infinite Series 255

Part V: The Part of Tens...**263**

Chapter 14: Ten Things about Limits, Continuity, and Infinite Series **265**

The 33333 Mnemonic...265
    First 3 over the "l": 3 parts to the definition of a limit.......................265
    Fifth 3 over the "l": 3 cases where a limit fails to exist......................266
    Second 3 over the "i": 3 parts to the definition of continuity............266
    Fourth 3 over the "i": 3 cases where continuity fails to exist............266
    Third 3 over the "m": 3 cases where a derivative fails to exist..........266
The 13231 Mnemonic...267
    First 1: The nth term test of divergence...267
    Second 1: The nth term test of convergence for alternating series....267
    First 3: The three tests with names...267
    Second 3: The three comparison tests...267
    The 2 in the middle: The two "R" tests..267

Chapter 15: Ten Things You Better Remember about Differentiation **269**

The Difference Quotient...269
The First Derivative Is a Rate 269
The First Derivative Is a Slope..269
Extrema, Sign Changes, and the First Derivative 270
The Second Derivative and Concavity 270
Inflection Points and Sign Changes in the Second Derivative....................270
The Product Rule..270
The Quotient Rule...270
Linear Approximation...271
"PSST," Here's a Good Way to Remember the Derivatives of Trig Functions.................271

**Chapter 16: Ten Things to Remember about Integration
If You Know What's Good for You** **273**

The Trapezoid Rule...273
The Midpoint Rule 273
Simpson's Rule 273
The Indefinite Integral 274
The Fundamental Theorem of Calculus, Take 1 274
The Fundamental Theorem of Calculus, Take 2 274
The Definite Integral 274
A Rectangle's Height Equals Top Minus Bottom..274
Area Below the *x*-Axis Is Negative 275
Integrate in Chunks..275

Index **277**

Introduction

*I*f you've already bought this book or are thinking about buying it, it's probably too late — too late, that is, to change your mind and get the heck out of calculus. (If you've still got a chance to break free, get out and run for the hills!) Okay, so you're stuck with calculus; you're past the point of no return. Is there any hope? Of course! For starters, buy this gem of a book and my other classic, *Calculus For Dummies*. In both books, you find calculus explained in plain English with a minimum of technical jargon. *Calculus For Dummies* covers topics in greater depth. *Calculus Workbook For Dummies* gives you the opportunity to master the calculus topics you study in class or in *Calculus For Dummies* through a couple hundred practice problems that will leave you giddy with the joy of learning . . . or pulling your hair out.

In all seriousness, calculus is not nearly as difficult as you'd guess from its reputation. It's a logical extension of algebra and geometry, and many calculus topics can be easily understood when you see the algebra and geometry that underlie them.

It should go without saying that regardless of how well you think you understand calculus, you won't fully understand it until you get your hands dirty by actually doing problems. On that score, you've come to the right place.

About This Book

Calculus Workbook For Dummies, like *Calculus For Dummies*, is intended for three groups of readers: high school seniors or college students in their first calculus course, students who've taken calculus but who need a refresher to get ready for other pursuits, and adults of all ages who want to practice the concepts they learned in *Calculus For Dummies* or elsewhere.

Whenever possible, I bring the calculus here down to earth by showing its connections to basic algebra and geometry. Many calculus problems look harder than they actually are because they contain so many fancy, foreign-looking symbols. When you see that the problems aren't that different from related algebra and geometry problems, they become far less intimidating.

I supplement the problem explanations with tips, shortcuts, and mnemonic devices. Often, a simple tip or memory trick can make it much easier to learn and retain a new, difficult concept.

Conventions Used in This Book

This book uses certain conventions:

- Variables are in *italics*.
- Important math terms are often in *italics* and defined when necessary.
- In the solution section, I've given your eyes a rest and not bolded all the numbered steps as is typical in *For Dummies* books.
- Extra hard problems are marked with an asterisk. You may want to skip these if you're prone to cerebral hemorrhaging.

How to Use This Book

Like all *For Dummies* books, you can use this book as a reference. You don't need to read it cover to cover or work through all problems in order. You may need more practice in some areas than others, so you may choose to do only half of the practice problems in some sections, or none at all.

However, as you'd expect, the order of the topics in *Calculus Workbook For Dummies* follows the order of the traditional curriculum of a first-year calculus course. You can, therefore, go through the book in order, using it to supplement your coursework. If I do say so myself, I expect you'll find that many of the explanations, methods, strategies, and tips in this book will make problems you found difficult or confusing in class seem much easier.

Foolish Assumptions

Now that you know a bit about how I see calculus, here's what I'm assuming about you:

✔ You haven't forgotten all the algebra, geometry, and trigonometry you learned in high school. If you have, calculus will be *really* tough. Just about every single calculus problem involves algebra, a great many use trig, and quite a few use geometry. If you're really rusty, go back to these basics and do some brushing up. This book contains some practice problems to give you a little pre-calc refresher, and *Calculus For Dummies* has an excellent pre-calc review.

✔ You're willing to invest some time and effort in doing these practice problems. Like with anything, practice makes perfect, and, also like anything, practice sometimes involves struggle. But that's a good thing. Ideally, you should give these problems your best shot before you turn to the solutions. Reading through the solutions can be a good way to learn, but you'll usually learn more if you push yourself to solve the problems on your own — even if that means going down a few dead ends.

How This Book Is Organized

Like all *For Dummies* books, this one is divided into parts, the parts into chapters, and the chapters into topics. Remarkable!

Part 1: Pre-Calculus Review

Part I is a brief review of the algebra, geometry, functions, and trigonometry that you'll need for calculus. You simply can't do calculus without a working knowledge of algebra and functions because virtually every single calculus problem involves both of these pre-calc topics in some way or another. You might say that algebra is the language calculus is written in and that functions are the objects that calculus analyzes. Geometry and trig are not quite as critical because you could do some calculus without them, but a great number of calculus problems and topics involve geometry and trig. If your pre-calc is rusty, get out the Rust-Oleum.

Part II: Limits and Continuity

You can actually do most practical calculus problems without knowing much about limits and continuity. The calculus done by scientists, engineers, and economists involves differential and integral calculus (see Parts III and IV), not limits and continuity. But because *mathematicians* do care about limits and continuity and because they're the ones who write calculus texts and design calculus curricula, you have to learn these topics.

Obviously, I'm being a bit cynical here. Limits and continuity are sort of the logical scaffolding that holds calculus up, and, as such, they're topics worthy of your time and effort.

Part III: Differentiation

Differentiation and integration (Part IV) are the two big ideas in calculus. Differentiation is the study of the *derivative,* or slope, of functions: where the slope is positive, negative, or zero; where the slope has a minimum or maximum value; whether the slope is increasing or decreasing; how the slope of one function is related to the slope of another; and so on. In Part III, you get differentiation basics, differentiation rules, and techniques for analyzing the shape of curves, and solving problems with the derivative.

Part IV: Integration and Infinite Series

Like differentiation, "integration" is a fancy word for a simple idea: addition. Every integration problem involves addition in one way or another. What makes integration such a big deal is that it enables you to add up an infinite number of infinitely small amounts. Using the magic of limits, integration cuts up something (an area, a volume, the pressure on the wall of a tank, and so on) into infinitely small chunks and then adds up the chunks to arrive at the total. In Part IV, you work through integration basics, techniques for finding integrals, and problem solving with integration.

Infinite series is a fascinating topic full of bizarre, counter-intuitive results, like the infinitely long trumpet shape that has an infinite surface area but a finite volume! — hard to believe but true. Your task with infinite series problems is to decide whether the sum of an infinitely long list of numbers adds up to infinity (something that's easy to imagine) or to some ordinary, finite number (something many people find hard to imagine).

Part V: The Part of Tens

Here you get ten things you should know about limits and infinite series, ten things you should know about differentiation, and ten things you should know about integration. If you find yourself knowing no calculus with your calc final coming up in 24 hours (perhaps because you were listening to Marilyn Manson on your iPod during class and did all your assignments in a "study" group), turn to the Part of Tens and the Cheat Sheet. If you learn only this material — not an approach I'd recommend — you may actually be able to barely survive your exam.

Icons Used in This Book

The icons help you to quickly find some of the most critical ideas in the book.

Next to this icon are important pre-calc or calculus definitions, theorems, and so on.

This icon is next to — are you sitting down? — example problems.

The tip icon gives you shortcuts, memory devices, strategies, and so on.

Ignore these icons and you'll be doing lots of extra work and probably getting the wrong answer.

Where to Go from Here

You can go

- ✔ To Chapter 1 — or to whatever chapter you need to practice.
- ✔ To *Calculus For Dummies* for more in-depth explanations. Then, because after finishing it and this workbook your newly acquired calculus expertise will at least double or triple your sex appeal, pick up *French For Dummies* and *Wine For Dummies* to impress Nanette or Jéan Paul.
- ✔ With the flow.
- ✔ To the head of the class, of course.
- ✔ Nowhere. There's nowhere to go. After mastering calculus, your life is complete.

Part I
Pre-Calculus Review

In this part . . .

Most of mathematics is cumulative — you can't do calculus without a solid knowledge of pre-calc. Obviously, there's much more to pre-calc than what's covered in the two short chapters of Part I, but if you're up to speed with the concepts covered here, you're in pretty good shape to begin the study of calculus. You really should be very comfortable with all this material, so work through the practice problems in Chapters 1 and 2, and if you find yourself on shaky ground, go back to your old textbooks (assuming you didn't burn them) or to the thorough pre-calc review in *Calculus For Dummies* to fill in any gaps in your knowledge of algebra, geometry, functions, and trig. Now you finally have an answer to the question you asked during high school math classes: "When am I ever going to need this?" Unfortunately, now there's the new question: "When am I ever going to need calculus?"

Chapter 1

Getting Down the Basics:
Algebra and Geometry

In This Chapter
▶ Fussing with fractions
▶ Brushing up on basic algebra
▶ Getting square with geometry

1 know, I know. This is a *calculus* workbook, so what's with the algebra and geometry? Don't worry, I'm not going to waste too many precious pages with algebra and geometry, but these topics are essential for calculus. You can no more do calculus without algebra than you can write French poetry without French. And basic geometry (but not geometry proofs — hooray!) is critically important because much of calculus involves real-world problems that include angles, slopes, shapes, and so on. So in this chapter — and in Chapter 2 on functions and trigonometry — I give you some quick problems to help you brush up on your skills. If you've already got these topics down pat, skip on over to Chapter 3.

If you miss some questions and don't quite understand why, go back to your old textbooks or check out the great pre-calc review in *Calculus For Dummies*. Getting these basics down pat is really important.

Fraction Frustration

Many, many math students hate fractions. Maybe the concepts didn't completely click when they first learned them and so fractions then became a nagging frustration in every subsequent math course.

But you can't do calculus without a good grasp of fractions. For example, the very definition of the derivative is based on a fraction called the *difference quotient*. And, on top of that, the symbol for the derivative, $\frac{dy}{dx}$, is a fraction. So, if you're a bit rusty with fractions, get up to speed with the following problems ASAP — or else!

Q. Solve $\frac{a}{b} \cdot \frac{c}{d} = ?$

A. $\frac{ac}{bd}$ To multiply fractions, you multiply straight across. You *do not* cross-multiply!

Q. Solve $\frac{a}{b} \div \frac{c}{d} = ?$

A. $\frac{a}{b} \div \frac{c}{d} = \frac{a}{b} \cdot \frac{d}{c} = \frac{ad}{bc}$ To divide fractions, you flip the second one, then multiply.

1. Solve $\frac{5}{0}$ = ?.

Solve It

2. Solve $\frac{0}{10}$ = ?.

Solve It

3. Does $\frac{3a+b}{3a+c}$ equal $\frac{a+b}{a+c}$? Why or why not?

Solve It

4. Does $\frac{3a+b}{3a+c}$ equal $\frac{b}{c}$? Why or why not?

Solve It

5. Does $\frac{4ab}{4ac}$ equal $\frac{ab}{ac}$? Why or why not?

Solve It

6. Does $\frac{4ab}{4ac}$ equal $\frac{b}{c}$? Why or why not?

Solve It

Misc. Algebra: You Know, Like Miss South Carolina

This section gives you a quick review of algebra basics like factors, powers and roots, logarithms, and quadratics. You absolutely *must* know these basics.

Q. Factor $9x^4 - y^6$.

A. $9x^4 - y^6 = (3x^2 - y^3)(3x^2 + y^3)$ This is an example of the single most important factor pattern: $a^2 - b^2 = (a-b)(a+b)$. Make sure you know it!

Q. Rewrite $x^{2/5}$ without a fraction power.

A. $\sqrt[5]{x^2} = (\sqrt[5]{x})^2$ Don't forget how fraction powers work!

7. Rewrite x^{-3} without a negative power.

Solve It

8. Does $(abc)^4$ equal $a^4b^4c^4$? Why or why not?

Solve It

9. Does $(a+b+c)^4$ equal $a^4 + b^4 + c^4$? Why or why not?

Solve It

10. Rewrite $\sqrt[3]{\sqrt[4]{x}}$ with a single radical sign.

Solve It

11. Does $\sqrt{a^2 + b^2}$ equal $a + b$? Why or why not?

Solve It

12. Rewrite $\log_a b = c$ as an exponential equation.

Solve It

13. Rewrite $\log_c a - \log_c b$ with a single log.

Solve It

14. Rewrite $\log 5 + \log 200$ with a single log and then solve.

Solve It

15. If $5x^2 = 3x + 8$, solve for x with the quadratic formula.

Solve It

16. Solve $|3x + 2| > 14$.

Solve It

Geometry: When Am I Ever Going to Need It?

You can use calculus to solve many real-world problems that involve surfaces, volumes, and shapes, such as maximizing the volume of a cylindrical soup can or determining the stress along a cable hanging in a parabolic shape. So you've got to know the basic geometry formulas for length, area, and volume. You also need to know basic stuff like the Pythagorean Theorem, proportional shapes, and basic coordinate geometry like the distance formula.

Q. What's the area of the triangle in the following figure?

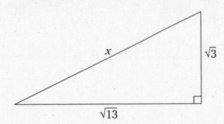

A. $\dfrac{\sqrt{39}}{2}$

$$Area_{triangle} = \frac{1}{2} \, base \cdot height$$
$$= \frac{1}{2} \cdot \sqrt{13}\sqrt{3}$$
$$= \frac{\sqrt{39}}{2}$$

Q. How long is the hypotenuse of the triangle in the previous example?

A. $x = 4$

$$a^2 + b^2 = c^2$$
$$x^2 = a^2 + b^2$$
$$x^2 = \sqrt{13}^2 + \sqrt{3}^2$$
$$x^2 = 13 + 3$$
$$x^2 = 16$$
$$x = 4$$

17. Fill in the two missing lengths for the sides of the triangle in the following figure.

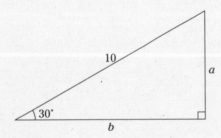

Solve It

18. What are the lengths of the two missing sides of the triangle in the following figure?

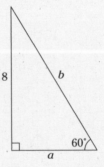

Solve It

19. Fill in the missing lengths for the sides of the triangle in the following figure.

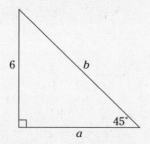

Solve It

20. a. What's the total area of the pentagon in the following figure?

b. What's the perimeter?

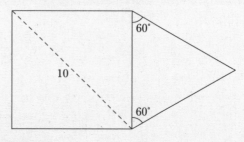

Solve It

21. Compute the area of the parallelogram in the following figure.

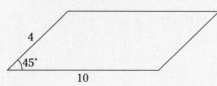

Solve It

22. What's the slope of $\overline{PQ}$?

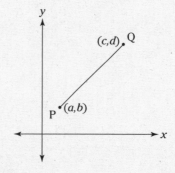

Solve It

23. How far is it from *P* to *Q* in the figure from problem 22?

Solve It

24. What are the coordinates of the midpoint of $\overline{PQ}$ in the figure from problem 22?

Solve It

Solutions for This Easy Elementary Stuff

1 Solve $\frac{5}{0} = ?$. $\frac{5}{0}$ **is undefined!** Don't mix this up with something like $\frac{0}{8}$, which equals zero. Note that if you think about these two fractions as examples of slope ($\frac{rise}{run}$), $\frac{5}{0}$ has a *rise* of 5 and a *run* of 0 which gives you a *vertical* line that has sort of an infinite steepness or slope (that's why it's undefined). Or just remember that it's impossible to drive up a vertical road and so it's impossible to come up with a slope for a vertical line. The fraction $\frac{0}{8}$, on the other hand, has a *rise* of 0 and a *run* of 8, which gives you a *horizontal* line that has no steepness at all and thus has the perfectly ordinary slope of zero. Of course, it's also perfectly ordinary to drive on a horizontal road.

2 $\frac{0}{10} = 0$ (See solution to problem 1.)

3 Does $\frac{3a+b}{3a+c}$ equal $\frac{a+b}{a+c}$? **No.** You can't cancel the 3s.

WARNING!

You can't cancel in a fraction unless there's an unbroken chain of multiplication running across the entire numerator and ditto for the denominator.

4 Does $\frac{3a+b}{3a+c}$ equal $\frac{b}{c}$? **No.** You can't cancel the 3*a*s. (See previous Warning.)

5 Does $\frac{4ab}{4ac}$ equal $\frac{ab}{ac}$? **Yes.** You can cancel the 4s because the entire numerator and the entire denominator are connected with multiplication.

6 Does $\frac{4ab}{4ac}$ equal $\frac{b}{c}$? **Yes.** You can cancel the 4*a*s.

7 Rewrite x^{-3} without a negative power. $\frac{1}{x^3}$

8 Does $(abc)^4$ equal $a^4 b^4 c^4$? **Yes.** Exponents do distribute over multiplication.

9 Does $(a+b+c)^4$ equal $a^4 + b^4 + c^4$? **No!** Exponents do *not* distribute over addition (or subtraction).

TIP

When you're working a problem and can't remember the algebra rule, try the problem with numbers instead of variables. Just replace the variables with simple, round numbers and work out the numerical problem. (Don't use 0, 1, or 2 because they have special properties that can mess up your example.) Whatever works for the numbers will work with variables, and whatever doesn't work with numbers won't work with variables. Watch what happens if you try this problem with numbers:

$$(3+4+6)^4 \overset{?}{=} 3^4 + 4^4 + 6^4$$
$$13^4 \overset{?}{=} 81 + 256 + 1296$$
$$28{,}561 \neq 1633$$

10 Rewrite $\sqrt[3]{\sqrt[4]{x}}$ with a single radical sign. $\sqrt[3]{\sqrt[4]{x}} = \sqrt[12]{x}$

11 Does $\sqrt{a^2 + b^2}$ equal $a + b$? **No!** The explanation is basically the same as for problem 9. Consider this: If you turn the root into a power, you get $\sqrt{a^2 + b^2} = (a^2 + b^2)^{1/2}$. But because you can't distribute the power, $(a^2 + b^2)^{1/2} \neq (a^2)^{1/2} + (b^2)^{1/2}$, or $a + b$, and thus $\sqrt{a^2 + b^2} \neq a + b$.

12 Rewrite $\log_a b = c$ as an exponential equation. $a^c = b$

13 Rewrite $\log_c a - \log_c b$ with a single log. $\log_c \frac{a}{b}$

14 Rewrite $\log 5 + \log 200$ with a single log and then solve. $\log 5 + \log 200 = \log(5 \cdot 200) = \log 1000 = \mathbf{3}$

When you see "log" without a base number, the base is 10.

15 If $5x^2 = 3x + 8$, solve for x with the quadratic formula. $x = \dfrac{8}{5}$ **or** -1

Start by rearranging $5x^2 = 3x + 8$ into $5x^2 - 3x - 8 = 0$ because you want just a zero on one side of the equation.

The quadratic formula tells you that $x = \dfrac{-b \pm \sqrt{b^2 - 4ac}}{2a}$. Plugging 5 into a, –3 into b, and –8

into c gives you $x = \dfrac{-(-3) \pm \sqrt{(-3)^2 - 4(5)(-8)}}{2 \cdot 5} = \dfrac{3 \pm \sqrt{9 + 160}}{10} = \dfrac{3 \pm 13}{10} = \dfrac{16}{10}$ or $\dfrac{-10}{10}$,

so $x = \dfrac{8}{5}$ or -1.

16 Solve $|3x + 2| > 14$. $x < -\dfrac{16}{3} \cup x > 4$

1. Turn the inequality into an equation: $|3x + 2| = 14$

2. Solve the absolute value equation.

$$3x + 2 = 14 \qquad\qquad 3x + 2 = -14$$
$$3x = 12 \quad \text{or} \quad 3x = -16$$
$$x = 4 \qquad\qquad\qquad x = -\frac{16}{3}$$

3. Place both solutions on a number line (see the following figure). (You use hollow dots for > and <; if the problem had been $\geq$ or $\leq$, you would use solid dots.)

4. Test a number from each of the three regions on the line in the original inequality.

For this problem you can use –10, 0, and 10.

$$\left|3 \cdot (-10) + 2\right| \overset{?}{>} 14$$
$$|-28| \overset{?}{>} 14$$
$$28 \overset{?}{>} 14$$

True, so you shade the left-most region.

$$\left|3 \cdot (0) + 2\right| \overset{?}{>} 14$$
$$2 \overset{?}{>} 14$$

False, so you don't shade the middle region.

$$|3 \cdot 10 + 2| \overset{?}{>} 14$$
$$|32| \overset{?}{>} 14$$
$$32 \overset{?}{>} 14$$

True, so shade the region on the right. The following figure shows the result. x can be any number where the line is shaded. That's your final answer.

5. If it floats your boat, you may also want to express the answer symbolically.

Because x can equal a number in the left region *or* a number in the right region, this is an *or* solution which means *union* ($\cup$). When you want to include everything from both regions on the number line, you want the *union* of the two regions. So, the symbolic answer is

$$x < -\frac{16}{3} \cup x > 4$$

If only the middle region were shaded, you'd have an *and* or *intersection* ($\cap$) problem. When you only want the section of the number line where the two regions overlap, you use the *intersection* of the two regions. Using the above number line points, for example, you would write the middle-region solution like

$$x > -\frac{16}{3} \text{ and } x < 4 \text{ or}$$

$$x > -\frac{16}{3} \cap x < 4 \text{ or}$$

$$-\frac{16}{3} < x < 4$$

You say "to-*may*-to," I say "to-*mah*-to."

While we're on the subject of absolute value, don't forget that $\sqrt{x^2} = |x|$. $\sqrt{x^2}$ does *not* equal $\pm x$.

17 Fill in the two missing lengths for the sides of the triangle. $a = 5$ and $b = 5\sqrt{3}$

This is a 30°-60°-90° triangle — Well, duhh!

18 Fill in the two missing lengths for the sides of the triangle.

$$a = \frac{8}{\sqrt{3}} \text{ or } \frac{8\sqrt{3}}{3}$$

$$b = \frac{16}{\sqrt{3}} \text{ or } \frac{16\sqrt{3}}{3}$$

Another 30°-60°-90° triangle.

19 Fill in the two missing lengths for the sides of the triangle. $a = 6$ and $b = 6\sqrt{2}$

Make sure you know your 45°-45°-90° triangle.

20 **a.** What's the total area of the pentagon? $50 + \dfrac{25\sqrt{3}}{2}$.

The square is $\frac{10}{\sqrt{2}}$ by $\frac{10}{\sqrt{2}}$ (because half a square is a 45°-45°-90° triangle), so the area is

$\frac{10}{\sqrt{2}} \cdot \frac{10}{\sqrt{2}} = \frac{100}{2} = 50$. The equilateral triangle has a base of $\frac{10}{\sqrt{2}}$, or $5\sqrt{2}$, so its height is $\frac{5\sqrt{6}}{2}$

(because half of an equilateral triangle is a 30°-60°-90° triangle). So the area of the triangle is

$\frac{1}{2}\left(5\sqrt{2}\right)\left(\frac{5\sqrt{6}}{2}\right) = \frac{25\sqrt{12}}{4} = \frac{50\sqrt{3}}{4} = \frac{25\sqrt{3}}{2}$. The total area is thus $50 + \frac{25\sqrt{3}}{2}$.

b. What's the perimeter? **The answer is $25\sqrt{2}$.**

The sides of the square are $\frac{10}{\sqrt{2}}$, or $5\sqrt{2}$, as are the sides of the equilateral triangle.

The pentagon has five sides, so the perimeter is $5 \cdot 5\sqrt{2}$, or $25\sqrt{2}$.

21 Compute the area of the parallelogram. **The answer is $20\sqrt{2}$.**

The height is $\dfrac{4}{\sqrt{2}}$, or $2\sqrt{2}$, because the height is one of the legs of a 45°-45°-90° triangle, and the base is 10. So, because the area of a parallelogram equals *base* times *height*, the area is $10 \cdot 2\sqrt{2}$, or $20\sqrt{2}$.

22 What's the slope of $\overline{PQ}$? $\dfrac{d-b}{c-a}$. Remember that $slope = \dfrac{rise}{run} = \dfrac{y_2 - y_1}{x_2 - x_1}$.

23 How far is it from P to Q? $\sqrt{(c-a)^2 + (d-b)^2}$

Remember that $distance = \sqrt{(x_2 - x_1)^2 + (y_2 - y_1)^2}$.

24 What are the coordinates of the midpoint of $\overline{PQ}$? $\left(\dfrac{a+c}{2}, \dfrac{b+d}{2}\right)$. The midpoint of a segment is given by the average of the two x coordinates and the average of the two y coordinates.

Chapter 2

Funky Functions and Tricky Trig

. . . .

In This Chapter

▶ Figuring functions

▶ Remembering Camp SohCahToa

. . . .

*I*n Chapter 2, you continue your pre-calc warm-up that you began in Chapter 1. If algebra is the language calculus is written in, you might think of functions as the "sentences" of calculus. And they're as important to calculus as sentences are to writing. You can't do calculus without functions. Trig is important not because it's an essential element of calculus — you could do most of calculus without trig — but because many calculus problems happen to involve trigonometry.

Figuring Out Your Functions

To make a long story short, a function is basically anything you can graph on your graphing calculator in " $y =$ " or graphing mode. The line $y = 3x - 2$ is a function, as is the parabola $y = 4x^2 - 3x + 6$. On the other hand, the sideways parabola $x = 3y^2 - 4y + 6$ isn't a function because there's no way to write it as $y = something$. Try it.

You can determine whether or not the graph of a curve is a function with the *vertical line test*. If there's no place on the graph where you could draw a vertical line that touches the curve more than once, then it *is* a function. And if you can draw a vertical line anywhere on the graph that touches the curve more than once, then it is *not* a function.

As you know, you can rewrite the above functions using " $f(x)$ " or " $g(x)$ " instead of " y ." This changes nothing; using something like $f(x)$ is just a convenient notation. Here's a sampling of calculus functions:

$$g'(x) = 3x^5 - 20x^3$$

$$f'(x) = \lim_{h \to 0} \frac{\sqrt{x+h} - \sqrt{x}}{h}$$

$$A_f(x) = \int_3^x 10dt$$

Virtually every single calculus problem involves functions in one way or another. So should you review some function basics? You betcha.

EXAMPLE

Q. If $f(x) = 3x^2 - 4x + 8$, what does $f(a+b)$ equal?

A. $3a^2 + 6ab + 3b^2 - 4a - 4b + 8$

$$f(x) = 3x^2 - 4x + 8$$
$$f(a+b) = 3(a+b)^2 - 4(a+b) + 8$$
$$= 3(a^2 + 2ab + b^2) - 4a - 4b + 8$$
$$= 3a^2 + 6ab + 3b^2 - 4a - 4b + 8$$

Q. For the line $g(x) = 5 - 4x$, what's the slope and what's the y-intercept?

A. **The slope is –4 and the y-intercept is 5.** Does $y = mx + b$ ring a bell? It better!

1. Which of the four relations shown in the figure represent functions and why? (A relation, by the way, is any collection of points on the x-y coordinate system.)

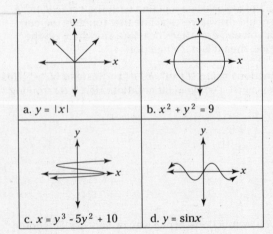

a. $y = |x|$

b. $x^2 + y^2 = 9$

c. $x = y^3 - 5y^2 + 10$

d. $y = \sin x$

Solve It

2. If the slope of line l is 3,

a. What's the slope of a line parallel to l?

b. What's the slope of a line perpendicular to l?

Solve It

3. Sketch a graph of $f(x) = e^x$.

Solve It

4. Sketch a graph of $g(x) = \ln x$.

Solve It

5. The following figure shows the graph of $f(x)$. Sketch the inverse of f, $f^{-1}(x)$.

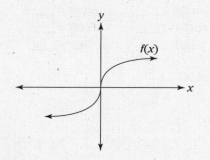

Solve It

6. The figure shows the graph of $p(x) = 2^x$. Sketch the following transformation of p: $q(x) = 2^{x+3} + 5$.

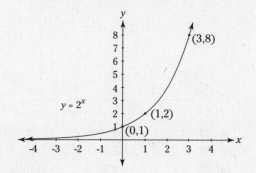

Solve It

Trigonometric Calisthenics

Believe it or not, trig is a very practical, real-world branch of mathematics, because it involves the measurement of lengths and angles. Surveyors use it when surveying property, making topographical maps, and so on. The ancient Greeks and Alexandrians, among others, knew not only simple SohCahToa stuff, but a lot of sophisticated trig as well. They used it for building, navigation, and astronomy. It's all over the place in the study of calculus, so if you snoozed through high school trig, WAKE UP! and review the following problems. (If you want to delve further into trig (and functions), check out *Calculus For Dummies*.)

7. Use the right triangle to complete the table.

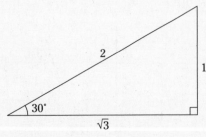

$\sin 30° =$ ____ $\csc 30° =$ ____
$\cos 30° =$ ____ $\sec 30° =$ ____
$\tan 30° =$ ____ $\cot 30° =$ ____

Solve It

8. Use the triangle from problem 7 to complete the following table.

$\sin 60° =$ ____ $\csc 60° =$ ____
$\cos 60° =$ ____ $\sec 60° =$ ____
$\tan 60° =$ ____ $\cot 60° =$ ____

Solve It

9. Use the following triangle to complete the table below.

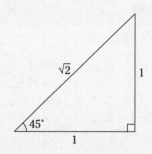

$\sin 45° =$ ____ $\csc 45° =$ ____
$\cos 45° =$ ____ $\sec 45° =$ ____
$\tan 45° =$ ____ $\cot 45° =$ ____

Solve It

10. Using your results from problems 7, 8, and 9, fill in the coordinates for the points on the unit circle.

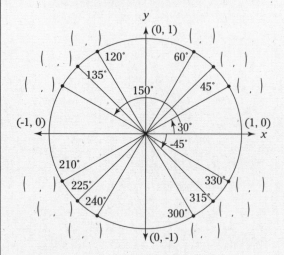

Solve It

11. Complete the following table using your results from problem 10.

$\tan 120° =$ ____ $\csc 180° =$ ____
$\csc 150° =$ ____ $\cot 300° =$ ____
$\cot 270° =$ ____ $\sec 225° =$ ____

Solve It

12. Convert the following angle measures from degrees to radians or vice-versa.

$150° =$ ____ radians
$225° =$ ____ radians
$300° =$ ____ radians
$-60° =$ ____ radians

$\dfrac{4\pi}{3} =$ ____ °

$\dfrac{7\pi}{4} =$ ____ °

$\dfrac{5\pi}{2} =$ ____ °

$-\dfrac{7\pi}{6} =$ ____ °

Solve It

13. Sketch $y = \sin x$ and $y = \cos x$.

Solve It

14. Using your answers from problem 10, complete the following table of inverse trig functions.

$$\sin^{-1}\left(\frac{1}{2}\right) = \underline{\quad}°$$

$$\sin^{-1}\left(-\frac{1}{2}\right) = \underline{\quad}°$$

$$\cos^{-1}\left(-\frac{1}{2}\right) = \underline{\quad}°$$

$$\tan^{-1}(-1) = \underline{\quad}°$$

$$\tan^{-1}\sqrt{3} = \underline{\quad} \text{ radians}$$

$$\sin^{-1}1 = \underline{\quad} \text{ radians}$$

$$\cos^{-1}1 = \underline{\quad} \text{ radians}$$

$$\cos^{-1}0 = \underline{\quad} \text{ radians}$$

Solve It

Solutions to Functions and Trigonometry

1 Which of the four relations in the figure represent functions and why? **A and D.**

The circle and the *S*-shaped curve are *not* functions because they fail the vertical line test: You *can* draw a vertical line somewhere on their graphs that touches the curve more than once. These two curves also fail the algebraic test: A curve is a function if for each input value (*x*) there is at most one output value (*y*). The circle and the *S*-shaped curve have some *x*s that correspond to more than one *y* so they are not functions. Note that the reverse is *not* true: You *can* have a function where there are two or more input values (*x*s) for a single output value (*y*).

2 If the slope of line *l* is 3,

   a. What's the slope of a line parallel to *l*? **The answer is 3.**

   b. What's the slope of a line perpendicular to *l*? **The answer is** $-\frac{1}{3}$**, the** *opposite reciprocal* **of 3.**

3 Sketch a graph of $f(x) = e^x$.

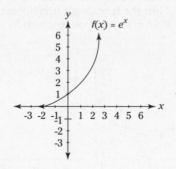

4 Sketch a graph of $g(x) = \ln x$.

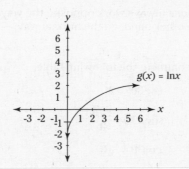

5 The figure shows the graph of $f(x)$. Sketch the inverse of f, $f^{-1}(x)$.

You obtain $f^{-1}(x)$ by reflecting $f(x)$ over the line $y = x$. See the figure.

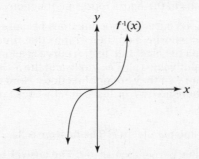

6 The figure shows the graph of $p(x) = 2^x$. Sketch the transformation of p, $q(x) = 2^{x+3} + 5$.

You obtain $q(x)$ from $p(x)$ by taking $p(x)$ and sliding it 3 to the *left* and 5 up. See the figure. Note that $q(x)$ contains "*x plus* 3," but the horizontal transformation is 3 to the *left* — the opposite of what you'd expect. The " + 5" in $q(x)$ tells you to go up 5.

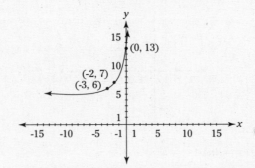

Horizontal transformations always work opposite the way you'd expect. Vertical transformations, on the other hand, go the normal way — *up* for *plus* and *down* for *minus*.

7 Use the right triangle to complete the following table.

$$\sin 30° = \frac{1}{2} \qquad\qquad \csc 30° = 2$$

$$\cos 30° = \frac{\sqrt{3}}{2} \qquad\qquad \sec 30° = \frac{2\sqrt{3}}{3}$$

$$\tan 30° = \frac{\sqrt{3}}{3} \qquad\qquad \cot 30° = \sqrt{3}$$

8 Use the triangle from problem 7 to complete the following table.

$$\sin 60° = \frac{\sqrt{3}}{2} \qquad\qquad \csc 60° = \frac{2\sqrt{3}}{3}$$

$$\cos 60° = \frac{1}{2} \qquad\qquad \sec 60° = 2$$

$$\tan 60° = \sqrt{3} \qquad\qquad \cot 60° = \frac{\sqrt{3}}{3}$$

9 Use the triangle to complete the table below.

$$\sin 45° = \frac{\sqrt{2}}{2} \qquad\qquad \csc 45° = \sqrt{2}$$

$$\cos 45° = \frac{\sqrt{2}}{2} \qquad\qquad \sec 45° = \sqrt{2}$$

$$\tan 45° = 1 \qquad\qquad \cot 45° = 1$$

10 Using your results from problems 7, 8, and 9, fill in the coordinates for the points on the unit circle.

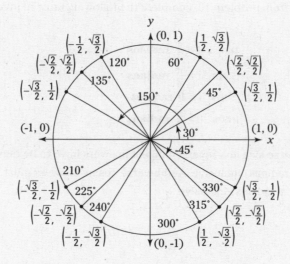

11 Complete the following table using your results from problem 10.

$$\tan 120° = -\sqrt{3} \qquad\qquad \csc 180° = \textbf{undefined}$$

$$\csc 150° = 2 \qquad\qquad \cot 300° = -\frac{\sqrt{3}}{3}$$

$$\sec 225° = -\sqrt{2}$$

$$\cot 270° = 0$$

12 Convert the following angle measures from degrees to radians or vice versa.

$$150° = \frac{5\pi}{6} \textbf{ radians} \qquad\qquad \frac{4\pi}{3} = \textbf{240}°$$

$$225° = \frac{5\pi}{4} \textbf{ radians} \qquad\qquad \frac{7\pi}{4} = \textbf{315}°$$

$$300° = \frac{5\pi}{3} \textbf{ radians} \qquad\qquad \frac{5\pi}{2} = \textbf{450}° \textbf{ (coterminal with 90}°\textbf{)}$$

$$-60° = -\frac{\pi}{3} \textbf{ radians} \qquad\qquad -\frac{7\pi}{6} = -\textbf{210}° \textbf{ (coterminal with 150}°\textbf{)}$$

13 Sketch $y = \sin x$ and $y = \cos x$.

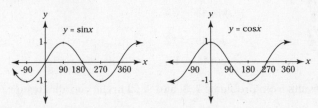

14 Using your answers from problem 10, complete the following table of inverse trigonometric functions.

$$\sin^{-1}\left(\frac{1}{2}\right) = \mathbf{30°} \qquad\qquad \tan^{-1}\sqrt{3} = \frac{\pi}{3} \textbf{ radians}$$

$$\sin^{-1}\left(-\frac{1}{2}\right) = \mathbf{-30°} \qquad\quad \sin^{-1}1 = \frac{\pi}{2} \textbf{ radians}$$

$$\cos^{-1}\left(-\frac{1}{2}\right) = \mathbf{120°} \qquad\quad \cos^{-1}1 = \mathbf{0 \textbf{ radians}}$$

$$\tan^{-1}(-1) = \mathbf{-45°} \qquad\qquad \cos^{-1}0 = \frac{\pi}{2} \textbf{ radians}$$

Don't forget — inverse sine and inverse tangent answers have to be between –90° and 90° (or $-\frac{\pi}{2}$ and $\frac{\pi}{2}$ radians) inclusive. And inverse cosine answers must be between 0° and 180° (or 0 and π radians) inclusive.

Part II
Limits and Continuity

"What exactly are we saying here?"

In this part . . .

Isaac Newton and Gottfried Leibniz, co-inventors of calculus in the late 17th century, did calculus without the solid foundation provided by limits and continuity. Now, since Newton and Leibniz were merely two of the greatest geniuses who ever lived, you don't want to do math at their lowly level, do you? Of course not! You want your calculus to be mathematically rigorous, and for that you need to master limits and continuity — despite their almost complete lack of practical importance. Now, don't blame me for this — heck, if practical relevance is your standard, you might as well drop out of school all together. Actually, the study of limits (and to a much lesser extent continuity) can be fascinating. And in the event that you hail from the show-me state, limits will allow you to prove that in a race against a tortoise where you give the tortoise a head start, you actually will catch up to and then pass the tortoise. The things that mathematics allows you to accomplish!

Chapter 3

A Graph Is Worth a Thousand Words: Limits and Continuity

· ·

In This Chapter

▶ The mathematical mumbo jumbo of limits and continuity

▶ When limits exist and *don't* exist

▶ Discontinuity. . . or graphus interruptus

· ·

You can use ordinary algebra and geometry when the things in a math problem *aren't* changing (sort of) and when lines are *straight*. But you need calculus when things *are* changing (these changing things are often represented as *curves*). For example, you need calculus to analyze something like the motion of the space shuttle during the beginning of its flight because its acceleration is changing every split second.

Ordinary algebra and geometry fall short for such things because the algebra or geometry formula that works one moment no longer works a millionth of a second later. Calculus, on the other hand, chops up these constantly changing things — like the motion of the space shuttle — into such tiny bits (actually infinitely small bits) that within each bit, things don't change. Then you *can* use ordinary algebra and geometry.

Limits are the "magical" trick or tool that does this chopping up of something into infinitely small bits. It's the mathematics of limits that makes calculus work. Limits are so essential that the formal definitions of the derivative and the definite integral both involve limits.

If — when your parents or other adults asked you, "What do you want to be when you grow up?" — you responded, "Why, a mathematician, of course," then you may ultimately spend a great deal of time thoroughly studying the deep and rich subtleties of *continuity*. For the rest of you, the concept of continuity is a total no-brainer. If you can draw a graph without lifting your pen or pencil from the page, the graph is *continuous*. If you can't — because there's a break in the graph — then the graph is not continuous. That's all there is to it. By the way, there are some subtle and technical connections between limits and continuity (which I don't want to get into), and that's why they're in the same chapter. But, be honest now, did you buy this book because you were dying to learn about mathematical subtleties and technicalities?

Digesting the Definitions: Limit and Continuity

This short section covers a couple formal definitions and a couple other things you need to know about limits and continuity. Here's the formal, three-part definition of a limit:

For a function $f(x)$ and a real number a, $\lim\limits_{x \to a} f(x)$ exists if and only if

1. $\lim\limits_{x \to a^-} f(x)$ exists. In other words, there must be a limit from the left.
2. $\lim\limits_{x \to a^+} f(x)$ exists. There must be a limit from the right.
3. $\lim\limits_{x \to a^-} f(x) = \lim\limits_{x \to a^+} f(x)$ The limit from the left must equal the limit from the right.

(Note that this definition does not apply to limits as x approaches infinity or negative infinity.)

And here's the definition of continuity: A function $f(x)$ is continuous at a point $x = a$ if three conditions are satisfied:

1. $f(a)$ is defined.
2. $\lim\limits_{x \to a} f(x)$ exists.
3. $f(a) = \lim\limits_{x \to a} f(x)$.

Using these definitions and Figure 3-1, answer problems 1 through 4.

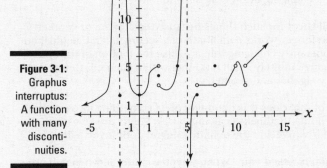

Figure 3-1:
Graphus
interruptus:
A function
with many
disconti-
nuities.

1. At which of the following *x*-values are all three requirements for the existence of a limit satisfied, and what is the limit at those *x*-values? $x = -2, 0, 2, 4, 5, 6, 8, 10,$ and 11.

Solve It

2. For the *x*-values at which all three limit requirements are not met, state which of the three requirements are not satisfied. If one or both one-sided limits exist at any of these *x*-values, give the value of the one-sided limit.

Solve It

3. At which of the *x*-values are all three requirements for continuity satisfied?

Solve It

4. For the rest of the *x*-values, state which of the three continuity requirements are not satisfied.

Solve It

Taking a Closer Look: Limit and Continuity Graphs

In this section, you get more practice at solving limit and continuity problems visually. Then in Chapter 4, you solve limit problems numerically (with your calculator) and symbolically (with algebra).

Use Figure 3-2 to answer problems 5 through 10.

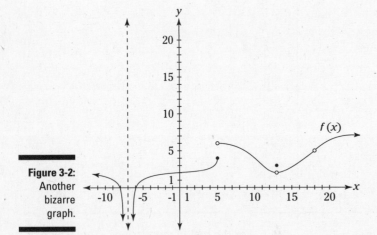

Figure 3-2:
Another
bizarre
graph.

Q. $\lim_{x \to 0} f(x) = ?$

A. $\lim_{x \to 0} f(x) = 2$ Because $f(0) = 2$ and because f is continuous there, the limit must equal the function value. Whenever a function passes through a point and there's no discontinuity at the point, the limit equals the function value.

Q. $\lim_{x \to 13} f(x) = ?$

A. $\lim_{x \to 13} f(x) = 2$ because there's a hole at $(13, 2)$. The limit at a hole is the height of the hole.

5. $\lim_{x \to -7} f(x) = ?$

Solve It

6. $\lim_{x \to 5} f(x) = ?$

Solve It

7. $\lim_{x \to 18} f(x) = ?$

Solve It

8. $\lim_{x \to 5^-} f(x) = ?$

Solve It

9. $\lim\limits_{x \to 5^+} f(x) = ?$

Solve It

10. List the *x*-coordinates of all points of discontinuity of *f* and state the type of discontinuity — removable, jump, or infinite.

Solve It

11. $\lim\limits_{x \to \infty} \sin x = ?$ See the following graph.

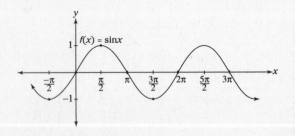

Solve It

12. $\lim\limits_{x \to \infty} \frac{1}{x} = ?$ See the following graph of $y = \frac{1}{x}$.

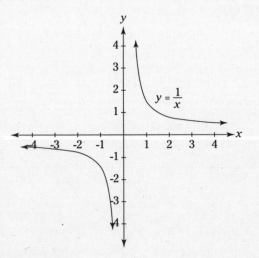

Solve It

Solutions for Limits and Continuity

1 At which of the following x-values are all three requirements for the existence of a limit satisfied, and what is the limit at those x-values? $x = -2, 0, 2, 4, 5, 6, 8, 10,$ and 11.

At 0, the limit is 2.

At 4, the limit is 5.

At 8, the limit is 3.

At 10, the limit is 5.

 REMEMBER

To make a long story short, a limit exists at a particular x-value of a curve when the curve is *heading toward* an exact y-value and keeps *heading toward* that y-value as you continue to zoom in on the curve at the x-value. The curve must head toward that y-value from the right and from the left (unless the limit is one where x approaches infinity). I emphasize "heading toward" because what happens precisely at the given x-value isn't relevant to this limit inquiry. That's why there is a limit at a hole like the ones at $x = 8$ and 10.

2 For the rest of the x-values, state which of the three limit requirements are not satisfied. If one or both one-sided limits exist at any of these x-values, give the value of the one-sided limit.

At −2 and 5, all three conditions fail.

At 2, 6, and 11, only the third requirement is not satisfied.

At 2, the limit from the left equals 5 and the limit from the right equals 3.

At 6, the limit from the left is 2 and the limit from the right is 3.

Finally, $\lim_{x \to 11^-} f(x) = 3$ and $\lim_{x \to 11^+} f(x) = 5$.

3 At which of the x-values are all three requirements for continuity satisfied?

The function in Figure 3-1 is continuous at 0 and 4. The common-sense way of thinking about continuity is that a curve is continuous wherever you can draw the curve without taking your pen off the paper. It should be obvious that that's true at 0 and 4, but not at any of the other listed x-values.

4 For the rest of the x-values, state which of the three continuity requirements are not satisfied.

All listed x-values other than 0 and 4 are points of discontinuity. A *discontinuity* is just a high-falutin' calculus way of saying a gap. If you'd have to take your pen off the paper at some point when drawing a curve, then the curve has a discontinuity there.

At 5 and 11, all three conditions fail.

At −2, 2, and 6, continuity requirements 2 and 3 are not satisfied.

At 10, requirements 1 and 3 are not satisfied.

At 8, requirement 3 is not satisfied.

5 $\lim_{x \to -7} f(x)$ **does not exist (DNE) because there's a vertical asymptote at −7.** Or, because $f(x)$ approaches $-\infty$ from the left and from the right, you could say the limit equals $-\infty$.

6 $\lim_{x \to 5} f(x)$ **does not exist because the limit from the left does not equal the limit from the right.** Or you could say that the limit DNE because there's a jump discontinuity at $x = 5$.

7 $\lim_{x \to 18} f(x) = 5$ **because, like the second example problem, the limit at a hole is the height of the hole.** The fact that $f(18)$ is undefined is irrelevant to this limit question.

8 $\lim_{x \to 5^-} f(x) = 4$ **because** $f(5) = 4$ **and** f **is continuous from the left at** $(5, 4)$.

9 $\lim_{x \to 5^+} f(x) = 6$. This question is just like problem 8 except that there's a hollow dot — instead of a solid one — at $(5, 6)$. But the hollow dot at $(5, 6)$ is irrelevant to the limit question — just like with question 7 where the hole was irrelevant.

10 List the x-coordinates of all points of discontinuity of f and state whether the points of discontinuity are removable or non-removable, and state the type of discontinuity — removable, jump, or infinite.

At $x = -7$**, the vertical asymptote, there is a non-removable, infinite discontinuity.**

At $x = 5$ **there's a non-removable, jump discontinuity.**

At $x = 13$ **and** $x = 18$ **there are holes which are removable discontinuities.** Though infinitely small, these are nevertheless discontinuities. They're "removable" discontinuities because you can "fix" the function by plugging the holes.

11 $\lim_{x \to \infty} \sin x$ **does not exist.** There's no limit as x approaches infinity because the curve oscillates — it never settles down to one exact y-value. (The three-part definition of a limit does not apply to limits at infinity.)

12 $\lim_{x \to \infty} \frac{1}{x} = 0$. In contrast to $\sin x$, this function does hone in on a single value; as you go out further and further to the right, the function gets closer and closer to zero, so that's the limit.

Chapter 4

Nitty-Gritty Limit Problems

• •

In This Chapter

▶ Algebra, schmalgebra

▶ Calculators — taking the easy way out

▶ Making limit sandwiches

▶ Infinity — "Are we there yet?"

▶ Conjugate multiplication — sounds *R* rated, but it's strictly *PG*

• •

In this chapter, you practice two very different methods for solving limit problems: using algebra and using your calculator. Learning the algebraic techniques are valuable for two reasons. The first, *incredibly* important reason is that the mathematics involved in the algebraic methods is beautiful, pure, and rigorous; and, second — something so trivial that perhaps I shouldn't mention it — you'll be tested on it. Do I have my priorities straight or what? The calculator techniques are useful for several reasons: 1) You can solve some limit problems on your calculator that are either impossible or just very difficult to do with algebra, 2) You can check your algebraic answers with your calculator, and 3) Limit problems can be solved with a calculator when you're not required to show your work — like maybe on a multiple choice test.

But before we get to these two major techniques, how about a little rote learning. A few limits are a bit tricky to justify or prove, so to make life easier, simply commit them to memory. Here they are:

▶ $\lim\limits_{x \to a} c = c$

($y = c$ is a horizontal line, so the limit equals c regardless of the *arrow-number* — the constant after the arrow.)

▶ $\lim\limits_{x \to 0^+} \dfrac{1}{x} = \infty$

▶ $\lim\limits_{x \to 0^-} \dfrac{1}{x} = -\infty$

▶ $\lim\limits_{x \to \infty} \dfrac{1}{x} = 0$

▶ $\lim\limits_{x \to -\infty} \dfrac{1}{x} = 0$

▶ $\lim\limits_{x \to 0} \dfrac{\sin x}{x} = 1$

▶ $\lim\limits_{x \to 0} \dfrac{\cos x - 1}{x} = 0$

▶ $\lim\limits_{x \to \infty} \left(1 + \dfrac{1}{x}\right)^{x} = e$

Solving Limits with Algebra

You can solve limit problems with several algebraic techniques. But your first step should always be plugging the arrow-number into the limit expression. If you get a number, that's the answer. You're done. You're also done if plugging in the arrow-number gives you

▶ A number or infinity or negative infinity over zero, like $\dfrac{3}{0}$, or $\dfrac{\pm\infty}{0}$; in these cases the limit does not exist (DNE).

▶ Zero over infinity; the answer is zero.

When plugging in fails because it gives you $\frac{0}{0}$, you've got a real limit problem, and you have to convert the fraction into some expression where plugging in *does* work. Here are some algebraic methods you can try:

- ✔ FOILing
- ✔ Factoring
- ✔ Finding the least common denominator
- ✔ Canceling
- ✔ Simplification
- ✔ Conjugate multiplication

The following examples each use a different method to solve the limit.

Q. Evaluate $\lim\limits_{x \to 16} \dfrac{16 - x}{4 - \sqrt{x}}$.

A. The limit is 8.

1. **Try plugging 16 into x — no good.**

2. **Multiply numerator and denominator by the conjugate of $4 - \sqrt{x}$, namely $4 + \sqrt{x}$.**

   The conjugate of a two-term expression has a plus sign instead of a minus sign — or vice-versa.

   $$\lim\limits_{x \to 16} \frac{(16 - x)}{\left(4 - \sqrt{x}\right)} \cdot \frac{\left(4 + \sqrt{x}\right)}{\left(4 + \sqrt{x}\right)}$$

3. **FOIL the conjugates and simplify.**

   $$= \lim\limits_{x \to 16} \frac{(16 - x)\left(4 + \sqrt{x}\right)}{\left(4^2 - \sqrt{x}^2\right)}$$ Because, of course, $(a - b)(a + b) = a^2 - b^2$.

   $$= \lim\limits_{x \to 16} \frac{(16 - x)\left(4 + \sqrt{x}\right)}{(16 - x)}$$

   $$= \lim\limits_{x \to 16} \left(4 + \sqrt{x}\right)$$

4. **Now you can cancel and then plug in.**

   $$= 4 + \sqrt{16}$$
   $$= 8$$

Note that while plugging in did not work in Step 1, it did work in the final step. That's your goal: to change the original expression — usually by canceling — so that plugging in works.

Q. What's $\lim\limits_{x \to -2} \dfrac{x^2 - x - 6}{x^2 + x - 2}$?

A. The limit is $\dfrac{5}{3}$.

1. **Try plugging –2 into x — that gives you $\frac{0}{0}$, so on to plan B.**

2. **Factor and cancel.**

   $$= \lim\limits_{x \to -2} \frac{(x + 2)(x - 3)}{(x + 2)(x - 1)}$$

   $$= \lim\limits_{x \to -2} \frac{(x - 3)}{(x - 1)}$$

3. **Cancel now and plug in.**

   $$= \frac{-2 - 3}{-2 - 1}$$

   $$= \frac{-5}{-3}$$

   $$= \frac{5}{3}$$

1. $\displaystyle\lim_{x \to 3} \frac{x^2 - 9}{x - 3}$

Solve It

2. $\displaystyle\lim_{x \to 1} \frac{x - 1}{x^2 + x - 2}$

Solve It

3. $\displaystyle\lim_{x \to -2} \frac{x + 2}{x^3 + 8}$

Solve It

4. $\displaystyle\lim_{x \to 2} \frac{x^2 - 4}{4x^2 + 5x - 6}$

Solve It

5. $\lim\limits_{x \to 9} \dfrac{x - 9}{3 - \sqrt{x}}$

Solve It

6. $\lim\limits_{x \to 10} \dfrac{\sqrt{x - 5} - \sqrt{5}}{x - 10}$

Solve It

7. $\lim\limits_{x \to 0} \dfrac{\cos x - 1}{x}$

Solve It

8. $\lim\limits_{x \to 2} \dfrac{\dfrac{1}{x} - \dfrac{1}{2}}{x - 2}$

Solve It

9. $\displaystyle\lim_{x \to 0} \frac{x}{\frac{1}{6} + \frac{1}{x-6}}$

Solve It

10. $\displaystyle\lim_{x \to 0} \frac{\sin x}{x}$

Solve It

*11. $\displaystyle\lim_{x \to 0} \frac{x}{\sin 3x}$

Solve It

*12. $\displaystyle\lim_{x \to 0} \frac{x}{\tan x}$

Solve It

Pulling Out Your Calculator: Useful "Cheating"

Your calculator is a great tool for understanding limits. It can often give you a better feel for how a limit works than the algebraic techniques can. A limit problem asks you to determine what the *y*-value of a function is zeroing in on as the *x*-value approaches a particular number. With your calculator, you can actually witness the process and the result. You can solve a limit problem with you calculator in three different ways.

Method I. First, store a number into *x* that's extremely close to the arrow-number, enter the limit expression in the home screen, and hit *enter*. If you get a result really close to a round number, that's your answer — you're done. If you have any doubt about the answer, just store another number into *x* that's even closer to the arrow-number and hit *enter* again. This will likely give you a result even closer to the same round number — that's it, you've got it. This method can be the quickest, but it often doesn't give you a good feel for how the *y*-values zero in on the result. To get a better picture of this process, you can store three or four numbers into *x* (one after another), each a bit closer to the arrow-number, and look at the sequence of results.

Method II. Enter the limit expression in graphing or "*y* =" mode, go to *Table Setup*, set *Tblstart* to the arrow-number, and set Δ*Tbl* to something small like 0.01 or 0.001. When you look at the table, you'll often see the *y*-values getting closer and closer to the limit answer as *x* hones in on the arrow-number. If it's not clear what the *y*-values are approaching, try a smaller increment for the Δ*Tbl* number. This method often gives you a good feel for what's happening in a limit problem.

Method III. This method gives you the best *visual* understanding of how a limit works. Enter the limit expression in graphing or "*y* =" mode. (If you're using the second method, you may want to try this third method at the same time.) Next, graph the function, and then go into the *window* and tweak the *xmin*, *xmax*, *ymin*, and *ymax* settings, if necessary, so that the part of the function corresponding to the arrow-number is within the viewing window. Use the *trace* feature to trace along the function until you get *close* to the arrow-number. You can't trace exactly *onto* the arrow-number because there's a little hole in the function there, the height of which, by the way, is your answer. When you trace close to the arrow-number, the *y*-value will get close to the limit answer. Use the *ZoomBox* feature to draw a little box around the part of the graph containing the arrow-number and zoom in until you see that the *y*-values are getting very close to a round number — that's your answer.

Q. Evaluate $\lim\limits_{x \to 6} \dfrac{x^2 - 5x - 6}{\sin(x - 6)}$.

A. **The answer is 7.**

Method I.

   1. **Use the *STO* → button to store 6.01 into *x*.**

   2. **Enter** $\dfrac{x^2 - 5x - 6}{\sin(x - 6)}$ **on the home screen and hit *enter*. (Note: You must be in *radian* mode.)**

   This gives you a result of ~7.01, suggesting that the answer is 7.

   3. **Repeat Steps 1 and 2 with 6.001 stored into *x*.**

   This gives you a result of ~7.001.

   4. **Repeat Steps 1 and 2 with 6.0001 stored into *x*.**

   This gives you a result of ~7.0001. Because the results are obviously honing in on the round number of 7, that's your answer.

Method II.

   1. **Enter** $\dfrac{x^2 - 5x - 6}{\sin(x - 6)}$ **in graphing or "*y* =" mode.**

   2. **Go to *Table Setup* and set *tblStart* to the arrow-number, 6, and *ΔTbl* to 0.01.**

   3. **Go to the *Table* and you'll see the *y*-values getting closer and closer to 7 as you scroll toward *x* = 6 from above and below 6.**

So 7 is your answer.

Method III.

   1. **Enter** $\dfrac{x^2 - 5x - 6}{\sin(x - 6)}$ **in graphing mode again.**

   2. **Graph the function. For your first viewing, *ZoomStd, ZoomFit,* and *ZoomTrig* (for expressions containing trig functions) are good windows to try.**

   For this funny function, none of these three window options works very well, but *ZoomStd* is the best.

   3. **Trace close to *x* = 6 and you'll see that *y* is near 7. Use *ZoomBox* to draw a little box around the point $(6, 7)$ then hit *enter*.**

   4. **Trace near *x* = 6 on this zoomed-in graph until you get very near to *x* = 6.**

   5. **Repeat the *Zoombox* process maybe two more times and you should be able to trace extremely close to *x* = 6.**

   (When I did this, I could trace to *x* = 6.0000022, *y* = 7.0000023.) The answer is 7.

13. Use your calculator to evaluate $\lim\limits_{x \to -3} \dfrac{x^2 - 5x - 24}{x + 3}$. Try all three methods.

Solve It

14. Use your calculator to determine $\lim\limits_{x \to 0} \dfrac{\sin x}{\tan^{-1} x}$. Use all three methods.

Solve It

Making Yourself a Limit Sandwich

The *sandwich* or *squeeze* method is something you can try when you can't solve a limit problem with algebra. The basic idea is to find one function that's always greater than the limit function (at least near the arrow-number) and another function that's always less than the limit function. Both of your new functions must have the same limit as x approaches the arrow-number. Then, because the limit function is "sandwiched" between the other two, like salami between slices of bread, it must have that same limit as well. See Figure 4-1.

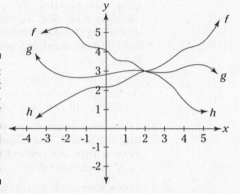

Figure 4-1:
A limit sandwich — functions f and h are the bread and g is the salami.

EXAMPLE

Q. What's $\lim\limits_{x \to 0} \dfrac{x}{\sqrt[3]{x}}$?

A. The limit is 0.

1. **Try plugging in 0. No good, you get 0 over 0.**

   You should be able to solve this limit problem with algebra. But let's say you tried and failed so now you're going to try the sandwich method.

2. **Graph the function.**

   Looks like the limit as x approaches 0 is 0.

3. **To prove it, try to find two bread functions that both have a limit of 0 as x approaches 0.**

   It's easy to show that the function is always positive (except perhaps at $x = 0$) so you can use the simple function $y = 0$ as the bottom slice of bread. Of course, it's obvious that $\lim\limits_{x \to 0} 0 = 0$. Finding a function for the top slice is harder. But let's just say that for some mysterious reason, you know that $y = \sqrt{|x|}$ is greater than $\dfrac{x}{\sqrt[3]{x}}$ near the arrow-number — the only place that matters for the sandwich method. Because $\lim\limits_{x \to 0} \sqrt{|x|} = 0$, $y = \sqrt{|x|}$ makes a good top slice.

   You're done. Because $\dfrac{x}{\sqrt[3]{x}}$ is squeezed between $y = 0$ and $y = \sqrt{|x|}$, both of which have limits of 0 as x approaches 0, $\dfrac{x}{\sqrt[3]{x}}$ must also have a limit of 0.

15. Evaluate $\lim\limits_{x \to 0}\left(x\sin\dfrac{1}{x^2} \right)$.

Solve It

16. Evaluate $\lim\limits_{x \to 0}\left(x^2\cos\dfrac{1}{x} \right)$.

Solve It

Into the Great Beyond: Limits at Infinity

To find a limit at infinity $\left(\lim\limits_{x \to \infty} \text{ or } \lim\limits_{x \to -\infty} \right)$, you can use the same techniques from the bulleted list in the "Solving Limits with Algebra" section of this chapter in order to change the limit expression so that you can plug in and solve.

If you're taking the limit at infinity of a *rational function* (which is one polynomial divided by another, such as $\dfrac{3x^2 - 8x + 12}{5x^3 + 4x^2 - x - 2}$), the limit will be the same as the *y*-value of the function's *horizontal asymptote,* which is an imaginary line that a curve gets closer and closer to as it goes right, left, up, or down toward infinity or negative infinity. Here are the two cases where this works:

✔ **Case 1:** If the degree of the polynomial in the numerator is *less than* the degree of the polynomial in the denominator, there's a horizontal asymptote at *y* = 0 and the limit as *x* approaches ∞ or −∞ is 0 as well.

✔ **Case 2:** If the degrees of the two polynomials are *equal,* there's a horizontal asymptote at the number you get when you divide the coefficient of the highest power term in the numerator by the coefficient of the highest power term in the denominator. This number is the answer to the limit as *x* approaches infinity or negative infinity. By the way, if the degree of the numerator is greater than the degree of the denominator, there's no horizontal asymptote and no limit.

Consider the following four types of expressions: x^{10}, 5^x, $x!$, and x^x. If a limit at infinity involves a fraction with one of them over another, you can apply a handy little tip. These four expressions are listed from "smallest" to "biggest." (This isn't a true ordering; it's only for problems of this type; and note that the actual numbers don't matter; they could just as easily be x^8, 3^x, $x!$, and x^x.) The limit will equal 0 if you have a "smaller" expression over a "bigger" one, and the limit will equal infinity if you have a "bigger" expression over a "smaller" one. And this rule is not affected by coefficients. For example, $\lim\limits_{x \to \infty} \dfrac{1000 \cdot x^{100}}{3x!} = 0$ and $\lim\limits_{x \to \infty} \dfrac{x^x}{500 \cdot 100^x} = \infty$. Note, however, that something like $(2x)!$ can change the ordering.

_Q._ Find $\lim\limits_{x \to \infty} \dfrac{x^3}{1.01^x}$.

_A._ **The limit is 0.**

This is an example of a "small" expression over a "big" one, so the answer is 0. Perhaps this result surprises you. You may think that this fraction will keep getting bigger and bigger because it seems that no matter what power 1.01 is raised to, it will never grow very large. And, in fact, if you plug 1000 into x, the quotient is big — over 47,000. But if you enter $\dfrac{x^3}{1.01^x}$ in graphing mode and then set both *tblStart* and Δtbl to 1000, the table values show quite convincingly that the limit is 0. By the time $x = 3000$, the answer is .00293, and when $x = 10{,}000$, the answer is 6×10^{-32}.

_Q._ $\lim\limits_{x \to \infty} \dfrac{100}{5x - \cos x^2}$

_A._ **The limit is 0.**

$$\lim_{x \to \infty} \frac{100}{5x - \cos x^2}$$
$$= \frac{100}{\infty}$$
$$= 0$$

The values of $\cos x^2$ that oscillate indefinitely between -1 and 1 are insignificant compared with $5x$ as x approaches infinity. Or consider the fact that $\lim\limits_{x \to \infty} \dfrac{100}{5x - 10} = 0$ and that $\dfrac{100}{5x - \cos x^2} < \dfrac{100}{5x - 10}$ for large values of x. Because $\dfrac{100}{5x - \cos x^2}$ is always positive for large values of x and less than something whose limit is 0, it must also have a limit of 0.

17. What's $\lim\limits_{x \to \infty} \dfrac{5x^3 - x^2 + 10}{2x^4 + x + 3}$? Explain your answer.

Solve It

18. What's $\lim\limits_{x \to -\infty} \dfrac{3x^4 + 100x^3 + 4}{8x^4 + 1}$? Explain your answer.

Solve It

19. Use your calculator to figure $\lim_{x \to \infty} \dfrac{x^x}{x!}$.

Solve It

20. Determine $\lim_{x \to \infty} \dfrac{5x+2}{\sqrt{4x^2-1}}$.

Solve It

*21. Evaluate $\lim_{x \to -\infty} \left(4x + \sqrt{16x^2 - 3x}\right)$.

Solve It

*22. Evaluate $\lim_{x \to -\infty} \left(\dfrac{3x^2}{x-1} - \dfrac{3x^2}{x+1}\right)$.

Solve It

Solutions for Problems with Limits

1 $\lim\limits_{x \to 3} \dfrac{x^2 - 9}{x - 3} = \mathbf{6}$

Factor, cancel, plug in.

$$= \lim_{x \to 3} \frac{(x - 3)(x + 3)}{(x - 3)}$$

$$= \lim_{x \to 3} \frac{x + 3}{1}$$

$$= \frac{3 + 3}{1}$$

$$= 6$$

2 $\lim\limits_{x \to 1} \dfrac{x - 1}{x^2 + x - 2} = \dfrac{\mathbf{1}}{\mathbf{3}}$

Factor, cancel, plug in.

$$= \lim_{x \to 1} \frac{(x - 1)}{(x - 1)(x + 2)}$$

$$= \lim_{x \to 1} \frac{1}{x + 2}$$

$$= \frac{1}{1 + 2}$$

$$= \frac{1}{3}$$

3 $\lim\limits_{x \to -2} \dfrac{x + 2}{x^3 + 8} = \dfrac{\mathbf{1}}{\mathbf{12}}$

Factor, cancel, plug in.

$$= \lim_{x \to -2} \frac{(x + 2)}{(x + 2)(x^2 - 2x + 4)}$$

$$= \lim_{x \to -2} \frac{1}{x^2 - 2x + 4}$$

$$= \frac{1}{(-2)^2 - 2(-2) + 4}$$

$$= \frac{1}{12}$$

4 $\lim\limits_{x \to 2} \dfrac{x^2 - 4}{4x^2 + 5x - 6} = \mathbf{0}$

Did you waste your time factoring the numerator and denominator? Gotcha! Always plug in first! When you plug 2 into the limit expression, you get $\frac{0}{20}$, or 0 — that's your answer.

5 $\lim\limits_{x \to 9} \dfrac{x - 9}{3 - \sqrt{x}} = \mathbf{-6}$

1. Multiply numerator and denominator by $3 + \sqrt{x}$.

$$= \lim_{x \to 9} \frac{(x - 9)}{(3 - \sqrt{x})} \cdot \frac{(3 + \sqrt{x})}{(3 + \sqrt{x})}$$

2. Multiply out the part of the fraction containing the conjugate pair (the denominator here).

$$=\lim_{x \to 9}\frac{(x-9)(3+\sqrt{x})}{(9-x)}$$

3. Cancel.

$$=\lim_{x \to 9}\left(-1(3+\sqrt{x})\right)$$

TIP

Don't forget that any fraction of the form $\frac{a-b}{b-a}$ always equals -1.

4. Plug in

$$= -1\left(3+\sqrt{9}\right)$$
$$= -6$$

6 $\lim_{x \to 10}\dfrac{\sqrt{x-5}-\sqrt{5}}{x-10} = \dfrac{\sqrt{5}}{10}$

Multiply by conjugate, multiply out, cancel, plug in.

$$=\lim_{x \to 10}\frac{\left(\sqrt{x-5}-\sqrt{5}\right)}{(x-10)} \cdot \frac{\left(\sqrt{x-5}+\sqrt{5}\right)}{\left(\sqrt{x-5}+\sqrt{5}\right)}$$

$$=\lim_{x \to 10}\frac{(x-5)-5}{(x-10)\left(\sqrt{x-5}+\sqrt{5}\right)}$$

$$=\lim_{x \to 10}\frac{(x-10)}{(x-10)\left(\sqrt{x-5}+\sqrt{5}\right)}$$

$$=\lim_{x \to 10}\frac{1}{\sqrt{x-5}+\sqrt{5}}$$

$$=\frac{1}{\sqrt{10-5}+\sqrt{5}}$$

$$=\frac{1}{2\sqrt{5}}$$

$$=\frac{\sqrt{5}}{10}$$

7 $\lim_{x \to 0}\dfrac{\cos x-1}{x} = 0$

Did you try multiplying the numerator and denominator by the conjugate of $\cos x - 1$? Gotcha again! That method doesn't work here. The answer to this limit is 0, something you just have to memorize.

8 $\lim_{x \to 2}\dfrac{\frac{1}{x}-\frac{1}{2}}{x-2} = -\dfrac{1}{4}$

1. Multiply numerator and denominator by the least common denominator of the little fractions inside the big fraction — that's $2x$.

$$=\lim_{x \to 2}\frac{\frac{1}{x}-\frac{1}{2}}{x-2} \cdot \frac{2x}{2x}$$

2. Multiply out the numerator.

$$=\lim_{x \to 2}\frac{(2-x)}{(x-2)(2x)}$$

3. Cancel.

$$= \lim_{x \to 2} \frac{-1}{2x}$$

4. Plug in.

$$= \frac{-1}{4}$$

$$= -\frac{1}{4}$$

9 $\lim\limits_{x \to 0} \dfrac{x}{\dfrac{1}{6} + \dfrac{1}{x-6}} = -36$

Multiply by the least common denominator, multiply out, cancel, plug in.

$$= \lim_{x \to 0} \frac{x}{\dfrac{1}{6} + \dfrac{1}{x-6}} \cdot \frac{6(x-6)}{6(x-6)}$$

$$= \lim_{x \to 0} \frac{6x(x-6)}{(x-6)+6}$$

$$= \lim_{x \to 0} \frac{6x(x-6)}{x}$$

$$= \lim_{x \to 0} 6(x-6)$$

$$= 6(0-6)$$

$$= -36$$

10 $\lim\limits_{x \to 0} \dfrac{\sin x}{x} = 1$

No work required — except for the memorization, that is.

*11 $\lim\limits_{x \to 0} \dfrac{x}{\sin 3x} = \dfrac{1}{3}$

Did you get it? If not, try the following hint before you read the solution: This fraction sort of resembles the one in problem 10. Still stuck? Okay, here you go:

1. Multiply numerator and denominator by 3.

You've got a $3x$ in the denominator, so you need $3x$ in the numerator as well (to make the fraction look more like the one in problem 10).

$$= \lim_{x \to 0} \frac{x}{\sin 3x} \cdot \frac{3}{3}$$

$$= \lim_{x \to 0} \frac{3x}{3 \sin 3x}$$

2. Pull the $\frac{1}{3}$ through the *lim* symbol (the 3 in the denominator is really a $\frac{1}{3}$, right?).

$$= \frac{1}{3} \lim_{x \to 0} \frac{3x}{\sin 3x}$$

Now, if your calc teacher lets you, you can just stop here — since it's "obvious" that $\lim\limits_{x \to 0} \dfrac{3x}{\sin 3x} = 1$ — and put down your final answer of $\frac{1}{3} \cdot 1$, or $\frac{1}{3}$. But if your teacher's a stickler for showing work, you'll have to do a couple more steps.

3. Set $u = 3x$.

4. Substitute u for $3x$. And, because u approaches 0 as x approaches 0, you can substitute u for x under the *lim* symbol.

$$=\frac{1}{3}\lim_{u \to 0}\frac{u}{\sin u}$$
$$=\frac{1}{3}\cdot 1$$
$$=\frac{1}{3}$$

Because $\lim\limits_{x \to 0}\frac{\sin x}{x} = 1$, the limit of the reciprocal of $\frac{\sin x}{x}$, namely $\frac{x}{\sin x}$, must equal the reciprocal of 1 — which is, of course, 1.

*12 $\lim\limits_{x \to 0}\frac{x}{\tan x} = 1$

1. Use the fact that $\lim\limits_{x \to 0}\frac{\sin x}{x} = 1$ and replace $\tan x$ with $\frac{\sin x}{\cos x}$.

$$=\lim_{x \to 0}\frac{x}{\frac{\sin x}{\cos x}}$$

2. Multiply numerator and denominator by $\cos x$.

$$=\lim_{x \to 0}\frac{x}{\frac{\sin x}{\cos x}}\cdot\frac{\cos x}{\cos x}$$
$$=\lim_{x \to 0}\frac{x\cos x}{\sin x}$$

3. Rewrite the expression as the product of two functions.

$$=\lim_{x \to 0}\left(\frac{x}{\sin x}\cdot\frac{\cos x}{1}\right)$$

4. Break this into two limits, using the fact that $\lim\limits_{x \to c}\big(f(x)\cdot g(x)\big)=\lim\limits_{x \to c}f(x)\cdot\lim\limits_{x \to c}g(x)$ (provided that both limits on the right exist).

$$=\lim_{x \to 0}\frac{x}{\sin x}\cdot\lim_{x \to 0}\cos x$$
$$=1\cdot 1 = 1$$

13 $\lim\limits_{x \to -3}\frac{x^2-5x-24}{x+3} = -11$

You want the limit as x approaches -3, so pick a number really close to -3 like -3.0001, plug that into x in your function $\frac{x^2-5x-24}{x+3}$ and enter that into your calculator. (If you've got a calculator like a Texas Instruments TI-83, TI-86, or TI-89, a good way to do this is to use the *STO*→ button to store -3.0001 into x, then enter $\frac{x^2-5x-24}{x+3}$ into the home screen and punch *enter*.) The calculator's answer is -11.0001. Because this is near the round number -11, your answer is -11. By the way, you can do this problem easily with algebra as well.

14 $\lim\limits_{x \to 0}\frac{\sin x}{\tan^{-1}x} = 1$

Enter the function in graphing mode like this: $y = \frac{\sin x}{\tan^{-1}x}$. Then go to *table setup* and enter a small increment into Δtbl (try 0.01 for this problem), and enter the arrow-number, 0, into *tblStart*. When you scroll through the table near $x = 0$, you'll see the y values getting closer and closer to the round number 1. That's your answer. This problem, unlike problem 13, is *not* easy to do with algebra.

15 Evaluate $\lim\limits_{x \to 0}\left(x\sin\dfrac{1}{x^2} \right) = \mathbf{0}$

Here are three ways to do this. First, common sense should tell you that this limit equals 0. $\lim\limits_{x \to 0} x$ is 0, of course, and $\lim\limits_{x \to 0}\left(\sin\dfrac{1}{x^2} \right)$ never gets bigger than 1 or smaller than –1. You could say that, therefore, $\lim\limits_{x \to 0}\left(\sin\dfrac{1}{x^2} \right)$ is "bounded" (bounded by –1 and 1). Then, because *zero* × *bounded* = *zero*, the limit is 0. Don't try this logic with you calc teacher — he won't like it.

Second, you can use your calculator: Store something small like .1 into x and then input $x\sin\dfrac{1}{x^2}$ into your home screen and hit *enter*. You should get a result of $\sim -.05$. Now store 0.01 into x and use the *entry* button to get back to $x\sin\dfrac{1}{x^2}$ and hit *enter* again. The result is $\sim .003$. Now try .001, then .0001 (giving you $\sim -.00035$ and $\sim .00009$) and so on. It's pretty clear — though probably not to the satisfaction of your professor — that the limit is 0.

The third way will definitely satisfy those typically persnickety professors. You've got to *sandwich* (or *squeeze*) your *salami* function, $x\sin\dfrac{1}{x^2}$, between two *bread* functions that have identical limits as x approaches the same arrow-number it approaches in the salami function. Because $\sin\dfrac{1}{x^2}$ never gets bigger than 1 or smaller than –1, $x\sin\dfrac{1}{x^2}$ will never get bigger than $|x|$ or smaller than $-|x|$. (You need the absolute value bars, by the way, to take care of negative values of x.) This suggests that you can use $b(x) = -|x|$ for the bottom piece of bread and $t(x) = |x|$ as the top piece of bread. Graph $b(x) = -|x|$, $f(x) = x\sin\dfrac{1}{x^2}$, and $t(x) = |x|$ at the same time on your graphing calculator and you can see that $x\sin\dfrac{1}{x^2}$ is always greater than or equal to $-|x|$ and always less than or equal to $|x|$. Because $\lim\limits_{x \to 0}(-|x|) = 0$ and $\lim\limits_{x \to 0}|x| = 0$ and because $x\sin\dfrac{1}{x^2}$ is sandwiched between them, $\lim\limits_{x \to 0}\left(x\sin\dfrac{1}{x^2} \right)$ must also be 0.

16 Evaluate $\lim\limits_{x \to 0}\left(x^2\cos\dfrac{1}{x} \right) = \mathbf{0}$

For $\lim\limits_{x \to 0}\left(x^2\cos\dfrac{1}{x} \right)$, use $b(x) = -x^2$ and $t(x) = x^2$ for the bread functions. The cosine of anything is always between –1 and 1, so $x^2\cos\dfrac{1}{x}$ is sandwiched between those two bread functions. (You should confirm this by looking at their graphs; use the following window on your graphing calculator — *Radian* mode, *xMin* = –0.15625, *xMax* = 0.15625, *xScl* = 0.05, *yMin* = –0.0125, *yMax* = 0.0125, *yScl* = 0.005.) Because $\lim\limits_{x \to 0}(-x^2) = 0$ and $\lim\limits_{x \to 0}x^2 = 0$, $\lim\limits_{x \to 0}\left(x^2\cos\dfrac{1}{x} \right)$ is also 0.

17 $\lim\limits_{x \to \infty}\dfrac{5x^3 - x^2 + 10}{2x^4 + x + 3} = \mathbf{0}$

Because the degree of the numerator is less than the degree of the denominator, this is a Case 1 problem. So the limit as x approaches infinity is 0.

18 $\lim\limits_{x \to -\infty}\dfrac{3x^4 + 100x^3 + 4}{8x^4 + 1} = \dfrac{\mathbf{3}}{\mathbf{8}}$

$\lim\limits_{x \to -\infty}\dfrac{3x^4 + 100x^3 + 4}{8x^4 + 1}$ is a Case 2 example because the degrees of the numerator and denominator are both 4. The limit is thus $\dfrac{3}{8}$.

19 $\lim\limits_{x \to \infty} \dfrac{x^x}{x!} = \infty$

According to the "big" over "small" tip, this answer must be infinity. Or you can get this result with your calculator. If you set the table (don't forget: fork on the left, spoon on the right) with something like *tblStart=100* and *Δtbl=100* and then look at the table, you may see "undef" for some or all of the *y* values depending on your calculator model. You've got to be careful when trying to interpret what "undef" (for "undefined") means on your calculator. It often means infinity but not always, so don't just jump to that conclusion. Instead, make *tblStart* and *Δtbl* smaller, say, 10. Sure enough, the *y* values grow huge very fast, and you can safely conclude that the limit is infinity.

20 $\lim\limits_{x \to \infty} \dfrac{5x + 2}{\sqrt{4x^2 - 1}} = \dfrac{5}{2}$

1. Divide numerator and denominator by x.

$$= \lim_{x \to \infty} \frac{\dfrac{5x + 2}{x}}{\dfrac{\sqrt{4x^2 - 1}}{x}}$$

2. Put the x into the square root (it becomes x^2).

$$= \lim_{x \to \infty} \frac{\dfrac{5x + 2}{x}}{\sqrt{\dfrac{4x^2 - 1}{x^2}}}$$

3. Distribute the division.

$$= \lim_{x \to \infty} \frac{5 + \dfrac{2}{x}}{\sqrt{4 - \dfrac{1}{x^2}}}$$

4. Plug in and simplify.

$$= \frac{5 + \dfrac{2}{\infty}}{\sqrt{4 - \dfrac{1}{\infty^2}}}$$

$$= \frac{5 + 0}{\sqrt{4 - 0}}$$

$$= \frac{5}{2}$$

*21 $\lim\limits_{x \to -\infty} \left(4x + \sqrt{16x^2 - 3x} \right) = \dfrac{3}{8}$

1. Put the entire expression over 1 so you can use the conjugate trick.

$$= \lim_{x \to -\infty} \frac{\left(4x + \sqrt{16x^2 - 3x} \right)}{1} \cdot \frac{\left(4x - \sqrt{16x^2 - 3x} \right)}{\left(4x - \sqrt{16x^2 - 3x} \right)}$$

2. FOIL the numerator.

$$= \lim_{x \to -\infty} \frac{16x^2 - \left(16x^2 - 3x \right)}{4x - \sqrt{16x^2 - 3x}}$$

3. Simplify the numerator and factor $16x^2$ inside the radicand.

$$= \lim_{x \to -\infty} \frac{3x}{4x - \sqrt{16x^2 \left(1 - \dfrac{3}{16x} \right)}}$$

4. Pull the $16x^2$ out of the square root — it becomes $-4x$.

It becomes *negative* $4x$ because x is negative when $x \to -\infty$, and because you've got to pull out a *positive*, you pull out $-4x$ because when x is negative, $-4x$ is positive. Got it?

$$= \lim_{x \to -\infty} \frac{3x}{4x - (-4x)\sqrt{1 - \dfrac{3}{16x}}}$$

$$= \lim_{x \to -\infty} \frac{3x}{4x\left(1 + \sqrt{1 - \dfrac{3}{16x}}\right)}$$

5. Cancel and plug in.

$$= \lim_{x \to -\infty} \frac{3}{4\left(1 + \sqrt{1 - \dfrac{3}{16x}}\right)}$$

$$= \frac{3}{4\left(1 + \sqrt{1 - \dfrac{3}{16(-\infty)}}\right)}$$

$$= \frac{3}{4\left(1 + \sqrt{1 - 0}\right)}$$

$$= \frac{3}{8} \qquad \text{Piece o' cake.}$$

*22 $\lim\limits_{x \to -\infty} \left(\dfrac{3x^2}{x-1} - \dfrac{3x^2}{x+1} \right) = 6$

1. Subtract the fractions using the LCD of $(x-1)(x+1) = x^2 - 1$.

$$\lim_{x \to \infty} \frac{3x^2(x+1) - 3x^2(x-1)}{x^2 - 1}$$

2. Simplify.

$$\lim_{x \to \infty} \frac{3x^3 + 3x^2 - 3x^3 + 3x^2}{x^2 - 1}$$

$$\lim_{x \to \infty} \frac{6x^2}{x^2 - 1}$$

3. Your answer is the quotient of the coefficients of x^2 in the numerator and the denominator (see Case 2 in the "Into the Great Beyond" section).

$$= 6$$

Note that had you plugged in ∞ in the original problem, you would have

$$\frac{3\infty^2}{\infty - 1} - \frac{3\infty^2}{\infty + 1}$$

$$= \infty - \infty$$

$$= 0?$$

It may seem strange, but infinity minus infinity does *not* equal 0.

Part III
Differentiation

The 5th Wave   By Rich Tennant

"I looked over your equations, Mrs. Dundt. Your concavity and inflection points are clean and there's nothing wrong with your velocity and acceleration. It might be your differentiation, but I won't be able to look at it until Thursday."

In this part . . .

In this part, you begin calculus in earnest. Differentiation is the first of the two big ideas in calculus (integration is the second). With all the mystique surrounding calculus, you may think that differentiation is as difficult a subject as string theory or ancient Sanskrit. But *differentiation* is simply the fancy calculus term for slope, which, as you remember from algebra I, is just the steepness of a line. Whenever possible, remind yourself that every problem involving differentiation is really just a slope problem. In Part III, you get differentiation basics, differentiation rules, techniques for using the derivative to analyze the shape of curves, and methods for solving practical problems with the derivative.

Chapter 5

Getting the Big Picture: Differentiation Basics

In This Chapter

▶ The ups and downs of finding slope and rate

▶ The difference quotient: the other DQ

Differentiation is the process of finding *derivatives*. The derivative is one of the most important inventions in the history of mathematics and one of mathematics' most powerful tools. I'm sure you will feel both a deep privilege as you do the practice problems below — and a keen sense of indebtedness to the great mathematicians of the past. Yeah, yeah, yeah.

The Derivative: A Fancy Calculus Word for Slope and Rate

A derivative of a function tells you how fast the output variable (like y) is changing compared to the input variable (like x). For example, if y is increasing 3 times as fast as x — like with the line $y = 3x + 5$ — then you say that the derivative of y with respect to x equals 3, and you write $\frac{dy}{dx} = 3$. This, of course, is the same as $\frac{dy}{dx} = \frac{3}{1}$, and that means nothing more than saying that the rate of change of y compared to x is 3-to-1, or that the line has a slope of $\frac{3}{1}$.

The following problems emphasize the fact that a derivative is basically just a rate or a slope. So to solve these problems, all you have to do is answer the questions as if they had asked you to determine a rate or a slope instead of a derivative.

Q. What's the derivative of $y = 4x - 5$?

A. **The answer is 4.** You know, of course, that the slope of $y = 4x - 5$ is 4, right? No? Egad! Any line of the form $y = mx + b$ has a slope equal to m. I hope that rings a bell. The derivative of a line or curve is the same thing as its slope, so the derivative of this line is 4.

You can think of the derivative $\frac{dy}{dx}$ as basically $\frac{rise}{run}$.

1. If you leave your home at time = 0, and go 60 $\frac{miles}{hour}$ in your car, what's $\frac{dp}{dt}$, the derivative of your position with respect to time?

Solve It

2. Using the information from problem 1, write a function that gives your position as a function of time.

Solve It

3. What's the slope of the parabola $y = -\frac{1}{3}x^2 + \frac{23}{3}x - \frac{85}{3}$ at the point (7, 9) in the following figure?

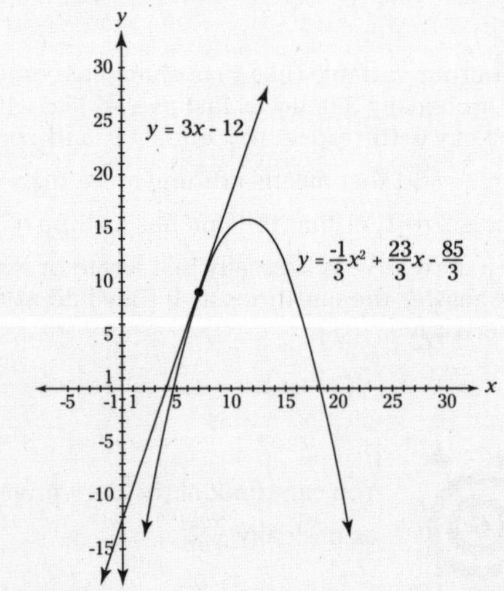

Solve It

4. What's the derivative of the parabola $y = -x^2 + 5$ at the point (0, 5)? Hint: Look at its graph.

Solve It

5. With your graphing calculator, graph both the line $y = -4x + 9$ and the parabola $y = 5 - x^2$. You'll see that they're tangent at the point (2, 1).

    a. What is the derivative of $y = 5 - x^2$ when $x = 2$?

    b. On the parabola, how fast is y changing compared to x when $x = 2$?

Solve It

6. Draw a function containing three points where — for three different reasons — you would not be able to determine the slope and, thus, where you would not be able to find a derivative.

Solve It

The Handy-Dandy Difference Quotient

The *difference quotient* is the almost magical tool that gives us the slope of a curve at a single point. To make a long story short, here's what happens when you use the difference quotient. (If you want an excellent version of the long story, check out *Calculus For Dummies*.) Look again at the figure in problem 3. You can't get the slope of the parabola at (7,9) with the algebra slope formula, $\left(m = \frac{y_2 - y_1}{x_2 - x_1} \right)$, because no matter what other point on the parabola you use with (7,9) in the formula, you'll get a slope that's steeper or less steep than the precise slope at (7,9).

But if your second point on the parabola is *extremely* close to (7,9) — like $\left(7.001, 9.002999\overline{6} \right)$ — your line would be almost exactly as steep as the tangent line. The difference quotient gives the precise slope of the tangent line by sliding the second point closer and closer to (7,9) until its distance from (7,9) is infinitely small.

Enough of this mumbo jumbo; now for the math. Here's the definition of the derivative based on the difference quotient:

$$f'(x) = \lim_{h \to 0} \frac{f(x+h) - f(x)}{h}$$

Like with most limit problems, plugging the arrow-number in at the beginning of a difference quotient problem won't help because that gives you $\frac{0}{0}$. You have to do a little algebraic mojo so that you can cancel the h and then plug in. (The techniques from Chapter 4 also work here.)

Now for a difference quotient problem.

Q. What's the slope of the parabola $f(x) = 10 - x^2$ at $x = 3$?

A. The slope is –6.

1. Because $f(x) = 10 - x^2$, then $f(x+h) = 10 - (x+h)^2$, and so the derivative is

$$f'(x) = \lim_{h \to 0} \frac{10 - (x+h)^2 - (10 - x^2)}{h}$$

2. Simplify.

$$= \lim_{h \to 0} \frac{10 - (x^2 + 2xh + h^2) - 10 + x^2}{h}$$

$$= \lim_{h \to 0} \frac{10 - x^2 - 2xh - h^2 - 10 + x^2}{h}$$

$$= \lim_{h \to 0} \frac{-2xh - h^2}{h}$$

3. Factor out h.

$$= \lim_{h \to 0} \frac{h(-2x - h)}{h}$$

4. Cancel.

$$= \lim_{h \to 0} (-2x - h)$$

5. Plug in the arrow-number.

$$= -2x - 0$$

$$f'(x) = -2x$$

6. You want the slope or derivative at $x = 3$, so plug in 3.

$$f'(3) = -2 \cdot 3$$

$$= -6$$

---

7. Use the difference quotient to determine the derivative of the line $y = 4x - 3$.

Solve It

8. Use the difference quotient to find the derivative of the parabola $f(x) = 3x^2$.

Solve It

9. Use the difference quotient to find the derivative of the parabola from problem 4, $y = -x^2 + 5$.

Solve It

10. Figure the derivative of $g(x) = \sqrt{4x+5}$ using the difference quotient.

Solve It

11. Use the parabola from problem 8, but make it a position function, $s(t) = 3t^2$, where t is in hours and $s(t)$ is in miles.

 a. What's the average velocity from $t = 4$ to $t = 5$?

 b. What's the average velocity from $t = 4$ to $t = 4.1$?

 c. What's the average velocity from $t = 4$ to $t = 4.01$?

Solve It

12. For the position function in problem 11, what's the *instantaneous* velocity at $t = 4$? *Hint:* Use the derivative.

Solve It

Solutions for Differentiation Basics

1 If you leave your home at time = 0, and go 60 $\frac{miles}{hour}$ in your car, what's $\frac{dp}{dt}$, the derivative of your position with respect to time? **The answer is $\frac{dp}{dt}$ = 60.** A derivative is always a rate, and a rate is always a derivative (assuming we're talking about instantaneous rates). So, if your speed, or rate, is 60 $\frac{miles}{hour}$, the derivative, $\frac{dp}{dt}$, is also 60. One way to think about a derivative like $\frac{dp}{dt}$ is that it tells you (in this case) how much your position (p) changes when the time (t) increases by one hour. A rate of 60 $\frac{miles}{hour}$ means that your position changes 60 miles each time the number of hours of your trip goes up by 1.

2 Using the information from problem 1, write a function that gives your position as a function of time. $p(t)$ = **60t or p = 60t, where t is in hours and p is in miles.** If you plug 1 into t, your position is 60 miles; plug 2 into t and your position is 120 miles. p = 60t is a line, of course, in the form $y = mx + b$ (where b = 0). So the slope is 60 and the derivative is thus 60. And again you see that a derivative is a slope and a rate.

3 What's the slope of the parabola $y = -\frac{1}{3}x^2 + \frac{23}{3}x - \frac{85}{3}$ at the point (7, 9)? **The slope is 3.**

You can see that the line, $y = 3x - 12$, is tangent to the parabola, $y = -\frac{1}{3}x^2 + \frac{23}{3}x - \frac{85}{3}$, at the point (7, 9). You know from $y = mx + b$ that the slope of $y = 3x - 12$ is 3. At the point (7, 9), the parabola is exactly as steep as the line, so the derivative (that's the slope) of the parabola at (7, 9) is also 3.

Although the slope of the line stays constant, the slope of the parabola changes as you climb up from (7, 9), getting less and less steep. Even if you go to the right just 0.001 to x = 7.001, the slope will no longer be exactly 3.

4 What's the derivative of the parabola $y = -x^2 + 5$ at the point (0, 5)? **The answer is 0.**

The point (0, 5) is the very top of the parabola, $y = -x^2 + 5$. At the top, the parabola is neither going up nor down — just like you're neither going up nor down when you're walking on top of a hill. The top of the parabola is flat or level in this sense, and thus the slope and derivative both equal zero.

The fact that the derivative is zero at the top of a hill — and at the bottom of a valley — is a critically important point which we'll return to time and time again.

5 With your graphing calculator, graph both the line $y = -4x + 9$ and the parabola $y = 5 - x^2$. You'll see that they're tangent at the point (2, 1).

a. What is the derivative of $y = 5 - x^2$ when x = 2? **The answer is –4.** The derivative of a curve tells you its slope or steepness. Because the line and the parabola are equally steep at (2, 1), and because you know the slope of the line is –4, the slope of the parabola at (2, 1) is also –4 and so is its derivative.

b. On the parabola, how fast is y changing compared to x when x = 2? **It's decreasing 4 times as fast as x increases.** A derivative is a rate as well as a slope. Because the derivative of the parabola is –4 at (2, 1), that tells you that y is changing 4 times as fast as x, but because the 4 is *negative*, y *decreases* 4 times as fast as x *increases*. This is the rate of y compared to x only for the one instant at (2, 1) — and thus it's called an *instantaneous* rate. A split second later — say at x = 2.000001 — y will be decreasing a bit faster.

6 Draw a function containing three points where — for three different reasons — you would not be able to determine the slope and thus where you would not be able to find a derivative.

Your sketch should contain (1) Any type of gap or discontinuity. There's no slope and thus no derivative at a gap because you can't draw a tangent line at a gap (try it). (2) A sharp point or cusp. It's impossible to draw a tangent line at a cusp because a line touching the function at such a sharp point could rock back and forth. So there's no slope and no derivative at a cusp. (3) A vertical inflection point. Although you *can* draw a tangent line at a vertical inflection point, since it's vertical, its slope — and therefore its derivative — is undefined.

7 Use the difference quotient to determine the derivative of the line $y = 4x - 3$. **$y' = 4$.**

$$y' = \lim_{h \to 0} \frac{4(x+h) - 3 - (4x - 3)}{h}$$
$$= \lim_{h \to 0} \frac{4x + 4h - 3 - 4x + 3}{h}$$
$$= \lim_{h \to 0} \frac{4h}{h}$$
$$= \lim_{h \to 0} 4$$
$$y' = 4$$

You can also figure this out because the slope of $y = 4x - 3$ is 4.

8 Use the difference quotient to find the derivative of the parabola $f(x) = 3x^2$. **$f'(x) = 6x$**

$$f'(x) = \lim_{h \to 0} \frac{3(x+h)^2 - 3x^2}{h}$$
$$= \lim_{h \to 0} \frac{3(x^2 + 2xh + h^2) - 3x^2}{h}$$
$$= \lim_{h \to 0} \frac{3x^2 + 6xh + 3h^2 - 3x^2}{h}$$
$$= \lim_{h \to 0} \frac{6xh + 3h^2}{h} \quad \text{(Now, factor out the } h\text{)}$$
$$= \lim_{h \to 0} \frac{h(6x + 3h)}{h} \quad \text{(Cancel the } h\text{)}$$
$$= \lim_{h \to 0} (6x + 3h) \quad \text{(Now plug in 0)}$$
$$= 6x + 3 \cdot 0$$
$$f'(x) = 6x$$

9 Use the difference quotient to find the derivative of the parabola from problem 4, $y = -x^2 + 5$. **$y' = -2x$.**

$$y' = \lim_{h \to 0} \frac{-(x+h)^2 + 5 - (-x^2 + 5)}{h}$$
$$= \lim_{h \to 0} \frac{-(x^2 + 2xh + h^2) + 5 + x^2 - 5}{h}$$
$$= \lim_{h \to 0} \frac{-x^2 - 2xh - h^2 + 5 + x^2 - 5}{h}$$
$$= \lim_{h \to 0} \frac{-2xh - h^2}{h} \quad \text{(Now factor)}$$
$$= \lim_{h \to 0} \frac{h(-2x - h)}{h} \quad \text{(And cancel)}$$
$$= \lim_{h \to 0} (-2x - h)$$
$$y' = -2x$$

In problem 4, you see that the top of this parabola ($y = -x^2 + 5$) is at the point (0,5) and that the derivative is zero there because the parabola is neither going up nor down at its peak. That explanation was based on common sense. But now, with the result given by the difference quotient, namely $y' = -2x$, you have a rigorous confirmation of the derivative's value at (0,5). Just plug 0 in for x in $y' = -2x$, and you get $y' = 0$.

10 Figure the derivative of $g(x) = \sqrt{4x+5}$ using the difference quotient. $g'(x) = \dfrac{2}{\sqrt{4x+5}}$

If you got this one, give yourself a pat on the back. It's a bit tricky.

$$g(x) = \sqrt{4x+5}$$

$$g'(x) = \lim_{h \to 0} \frac{\sqrt{4(x+h)+5} - \sqrt{4x+5}}{h}$$

$$= \lim_{h \to 0} \frac{\sqrt{4x+4h+5} - \sqrt{4x+5}}{h}$$

$$= \lim_{h \to 0} \frac{\left(\sqrt{4x+4h+5} - \sqrt{4x+5}\right)}{h} \cdot \frac{\left(\sqrt{4x+4h+5} + \sqrt{4x+5}\right)}{\left(\sqrt{4x+4h+5} + \sqrt{4x+5}\right)} \quad \text{(Conjugate multiplication)}$$

$$= \lim_{h \to 0} \frac{(4x+4h+5) - (4x+5)}{h\left(\sqrt{4x+4h+5} + \sqrt{4x+5}\right)} \quad \left(\text{Because } (a-b)(a+b) = a^2 - b^2, \text{ of course}\right)$$

$$= \lim_{h \to 0} \frac{4h}{h\left(\sqrt{4x+4h+5} + \sqrt{4x+5}\right)}$$

$$= \lim_{h \to 0} \frac{4}{\sqrt{4x+4h+5} + \sqrt{4x+5}} \quad \text{(Now you can plug in)}$$

$$= \frac{4}{\sqrt{4x+4 \cdot 0+5} + \sqrt{4x+5}}$$

$$= \frac{4}{2\sqrt{4x+5}}$$

$$g'(x) = \frac{2}{\sqrt{4x+5}}$$

11 Use the parabola from problem 8, but make it a position function, $s(t) = 3t^2$, where t is in hours and $s(t)$ is in miles.

Average velocity equals $\dfrac{total\ distance}{total\ time}$.

a. What's the average velocity from $t = 4$ to $t = 5$? **The answer is 27 *miles/hour.***

$$Average\ Velocity_{4\,to\,5} = \frac{s(5) - s(4)}{5 - 4}$$

$$= \frac{3 \cdot 5^2 - 3 \cdot 4^2}{1}$$

$$= 27 \frac{miles}{hour}$$

b. What's the average velocity from $t = 4$ to $t = 4.1$? **The answer is 24.3 *miles/hour.***

$$Average\ Velocity_{4\,to\,4.1} = \frac{s(4.1) - s(4)}{4.1 - 4}$$

$$= \frac{3 \cdot 4.1^2 - 3 \cdot 4^2}{0.1}$$

$$= 24.3 \frac{miles}{hour}$$

c. What's the average velocity from $t = 4$ to $t = 4.01$? **The answer is 24.03** *miles/hour.*

$$\text{Average Velocity}_{4 \text{ to } 4.01} = \frac{s(4.01) - s(4)}{4.01 - 4}$$

$$= \frac{3 \cdot 4.01^2 - 3 \cdot 4^2}{0.01}$$

$$= 24.03 \frac{miles}{hour}$$

12 For the position function in problem 11, what's the *instantaneous* velocity at $t = 4$? **The answer is 24** *miles/hour.* Problem 8 gives you the derivative of this parabola, $f'(x) = 6x$. The position function in this problem is the same except for different variables, so its derivative is $s'(t) = 6t$. Plug in 4, and you get $s'(4) = 24 \frac{miles}{hour}$. Notice how the average velocities get closer and closer to $24 \frac{miles}{hour}$ as the total travel time gets less and less and the ending time hones in on $t = 4$. That's precisely how the difference quotient works as h shrinks to zero.

Chapter 6

Rules, Rules, Rules: The Differentiation Handbook

. . . .

In This Chapter

▶ Boning up on basic derivative rules

▶ Producing your quota of product and quotient problems

▶ Joining the chain rule gang

▶ Achieving higher order differentiation

. . . .

Chapter 5 gives you the meaning of the derivative. In this chapter, you practice rules for finding derivatives. But before you practice the following rules, you may want to go back to the Cheat Sheet in *Calculus For Dummies* or to your calc text to review the basic derivatives. For example, you need to know that the derivative of sine is cosine.

Rules for Beginners

Okay, now that you've got the memorization stuff taken care, you now begin some rules that involve more than just memorizing the answer.

First there's the derivative of a constant rule: The derivative of a constant is zero. Alright — this one's also just memorization.

And then there's the power rule: To find the derivative of a variable raised to a power, bring the power in front — multiplying it by the coefficient, if there is one — then reduce the power by one.

Q. What's the derivative of $5x^3$?

A. $15x^2$

  1. Bring the power in front, multiplying it by the coefficient.

  So far you've got $15x^3$. Note that this does not equal $5x^3$ and so you should not put an equal sign in front of it.

In fact, there's no reason to write this interim step down at all. I do it simply to make the process clear.

  2. Reduce the power by one.

  This gives you the final answer of $15x^2$.

1. What's the derivative of $f(x) = 8$?

Solve It

2. What's the derivative of $g(x) = \pi^3$?

Solve It

3. What's the derivative of $g(x) = k \sin \frac{\pi}{2} \cos 2\pi$, where k is a constant?

Solve It

4. For $f(x) = 5x^4$, $f'(x) = ?$

Solve It

5. For $g(x) = \frac{-x^3}{10}$, what's $g'(x)$?

Solve It

6. Find y' if $y = \sqrt{x^{-5}}$ $(x \geq 0)$

Solve It

7. What's the derivative of $s(t) = 7t^6 + t + 10$?

Solve It

8. Find the derivative for $y = (x^3 - 6)^2$.

Solve It

Giving It Up for the Product and Quotient Rules

Now that you've got the easy stuff down, I'm sure you're dying to get some practice with advanced differentiation rules. The *product rule* and the *quotient rule* give you the derivatives for the product of two functions and the quotient of two functions, respectively and obviously.

The *product rule* is a snap. The derivative of a product of two functions, *(first)(second)*, equals $(\textit{first})'(\textit{second}) + (\textit{first})(\textit{second})'$.

The *quotient rule* is also a piece of cake. The derivative of a quotient of two functions, $\dfrac{(\textit{first})}{(\textit{second})}$, equals $\dfrac{(\textit{first})'(\textit{second}) - (\textit{first})(\textit{second})'}{(\textit{second})^2}$.

Here's a good way to remember the quotient rule. Notice that the numerator of the quotient rule looks exactly like the product rule, except that there's a minus sign instead of a plus sign. And note that when you read a product, you read from left to right, and when you read a quotient, you read from top to bottom. So just remember that the quotient rule, like the product rule, works in the natural order in which you read, beginning with the derivative of the first thing you read.

For some mysterious reason, many textbooks give the quotient rule in a different form that's harder to remember. Learn it the way I've written it above, beginning with $(\textit{first})'$. That's the easiest way to remember it.

Q. $\dfrac{d}{dx}\left(x^2 \sin x\right) = ?$

A. $\dfrac{d}{dx}\left(x^2 \sin x\right) = \left(x^2\right)'(\sin x) + \left(x^2\right)(\sin x)'$
$= 2x \sin x + x^2 \cos x$

Q. $\dfrac{d}{dx}\dfrac{x^2}{(\sin x)} = ?$

A. $\dfrac{d}{dx}\dfrac{x^2}{(\sin x)} = \dfrac{\left(x^2\right)'(\sin x) - \left(x^2\right)(\sin x)'}{(\sin x)^2}$
$= \dfrac{2x \sin x - x^2 \cos x}{\sin^2 x}$

One more thing: I've purposely designed this example to resemble the product rule example, so you can see the similarity between the quotient rule numerator and the product rule.

9. $\frac{d}{dx}\left(x^3 \cos x\right) = ?$

Solve It

10. $\frac{d}{dx}\left(\sin x \tan x\right) = ?$

Solve It

11. $\frac{d}{dx}\left(5x^3 \ln x\right) = ?$

Solve It

*12. $\frac{d}{dx}\left(x^2 e^x \ln x\right) = ?$

Solve It

13. $\frac{d}{dx}\frac{x^3}{\cos x} = ?$

Solve It

14. $\frac{d}{dx}\frac{\cos x}{e^x} = ?$

Solve It

15. $\frac{d}{dx}\frac{3x^2 + 3}{\arctan x} = ?$

Solve It

*16. $\frac{d}{dx}\frac{\sin x}{x^3 \ln x} = ?$

Solve It

Linking Up with the Chain Rule

The chain rule is probably the trickiest among the advanced rules, but it's really not that bad at all if you focus clearly on what's going on — I promise! Most of the basic derivative rules have a plain old x as the argument (or input variable) of the function. For example, $f(x) = \sqrt{x}$, $\sin x$, and $y = e^x$ all have just x as the argument.

When the argument of a function is anything other than a plain old x, such as $y = \sin x^2$ or $\ln 10^x$, you've got a chain rule problem.

Here's what you do. You simply apply the derivative rule that's appropriate to the outer function, temporarily ignoring the not-a-plain-old-x argument. Then multiply that result by the derivative of the argument. That's all there is to it.

Q. What's the derivative of $y = \sin x^3$?

A. $y' = 3x^2 \cos x^3$

1. **Temporarily think of the argument, x^3, as a _glob_.**

   So, you've got $y = \sin(glob)$.

2. **Use the regular derivative rule.**

   $y = \sin(glob)$, so

   $y' = \cos(glob)$

   (This is only a provisional answer, so the "=" sign is false — egad! The math police are going to pull me over.)

3. **Multiply this by the derivative of the argument.**

   $y' = \cos(glob) \cdot glob'$

4. **Get rid of the _glob_.**

   The _glob_ equals x^3 so _glob'_ equals $3x^2$.

   $y' = \cos(x^3) \cdot 3x^2$

   $= 3x^2 \cos x^3$

Q. What's the derivative of $\sin^4 x^3$? You've got to use the chain rule twice for this one.

A. The answer is $12x^2 \sin^3 x^3 \cdot \cos x^3$

1. **Rewrite $\sin^4 x^3$ to show what it really means: $\left(\sin x^3\right)^4$**

2. **The outermost function is the 4th power, so use the derivative rule for _stuff_⁴ — that's $4stuff^3$; then multiply that by the derivative of the inside _stuff_, $\sin x^3$.**

   $4\left(\sin x^3\right)^3 \cdot \left(\sin x^3\right)'$

   With chain rule problems, always work _from the outside, in_.

3. **To get the derivative of $\sin x^3$, use the derivative rule for $\sin(glob)$, and then multiply that by _glob'_.**

   $4\left(\sin x^3\right)^3 \cdot \cos(x^3) \cdot (x^3)'$

4. **The derivative of x^3 is $3x^2$, so you've got**

   $4\left(\sin x^3\right)^3 \cdot \cos(x^3) \cdot 3x^2$

5. **To simplify, rewrite the _sin_ power and move the $3x^2$ to the front.**

   $= 12x^2 \sin^3 x^3 \cdot \cos x^3$

   Can you remember when you've had so much fun?

   With chain rule problems, never use more than one derivative rule per step. In other words, when you do the derivative rule for the outermost function, don't touch the inside stuff! Only in the next step do you multiply the outside derivative by the derivative of the inside stuff.

17. $f(x) = \sin x^2$
$f'(x) = ?$

Solve It

18. $g(x) = \sin^3 x$
$g'(x) = ?$

Solve It

19. $s(t) = \tan(\ln t)$
$s'(t) = ?$

Solve It

20. $y = e^{4x^3}$
$y' = ?$

Solve It

*21. $f(x) = x^4 \sin^3 x$
$f'(x) = ?$

Solve It

*22. $g(x) = \dfrac{(\ln x)^2}{5x - 4}$
$g'(x) = ?$

Solve It

*23. $y = \cos^3 4x^2$
$y' = ?$

Solve It

*24. $\dfrac{d}{dx}\left(\tan^3 e^{x^2}\right) = ?$

Solve It

What to Do with Ys: Implicit Differentiation

You use implicit differentiation when your equation isn't in "$y =$" form, such as $\sin y^2 = x^3 + 5y^3$, and it's impossible to solve for y. If you can solve for y, implicit differentiation will still work, but it's not necessary.

Implicit differentiation problems are chain rule problems in disguise. Here's what I mean. You know that the derivative of $\sin x$ is $\cos x$, and that according to the chain rule, the derivative of $\sin x^3$ is $\cos(x^3) \cdot (x^3)'$. You would finish that problem by doing the derivative of x^3, but I have a reason for leaving the problem unfinished.

To do implicit differentiation, all you do (sort of) is every time you see a "y" in a problem, you treat it as if it were x^3. Thus, because the derivative of $\sin x^3$ is $\cos(x^3) \cdot (x^3)'$, the derivative of $\sin y$ is $\cos y \cdot y'$. Then, after doing the differentiation, you just rearrange terms so that you get $y' = something$.

 By the way, I used "y" in the explanation above, but that's not the whole story. Consider that $y' = 20x^3$ is the same as $\dfrac{dy}{dx} = 20x^3$. It's the variable on the top that you apply implicit differentiation to. This is typically y, but it could be any other variable. And it's the variable on the bottom that you treat the ordinary way. This is typically x, but it could also be any other variable.

Q. If $y^3 + x^3 = \sin y + \sin x$, find $\dfrac{dy}{dx}$.

A. $y' = \dfrac{\cos x - 3x^2}{3y^2 - \cos y}$

1. **Take the derivative of all four terms, using the chain rule for terms containing y and using the ordinary method for terms containing x.**

   $3y^2 \cdot y' + 3x^2 = \cos y \cdot y' + \cos x$

2. **Move all terms containing y' to the left side and all other terms to the right side.**

   $3y^2 \cdot y' - \cos y \cdot y' = \cos x - 3x^2$

3. **Factor out y'.**

   $y'(3y^2 - \cos y) = \cos x - 3x^2$

4. **Divide.**

   $y' = \dfrac{\cos x - 3x^2}{3y^2 - \cos y}$

   That's your answer. Note that this derivative — unlike ordinary derivatives — contains ys as well as xs.

25. If $y^3 - x^2 = x + y$, find $\dfrac{dy}{dx}$ by implicit differentiation.

Solve It

26. If $3y + \ln y = 4e^x$, find y'.

Solve It

27. For $x^2 y = y^3 x + 5y + x$, find $\dfrac{dy}{dx}$ by implicit differentiation.

Solve It

*28. If $y + \cos^2 y^3 = \sin 5x^2$, find the slope of the curve at $\left(\sqrt{\dfrac{\pi}{10}}, 0 \right)$.

Solve It

Getting High on Calculus: Higher Order Derivatives

You often need to take the derivative of a derivative, or the derivative of a derivative of a derivative, and so on. In the next two chapters, you see a few applications. For example, a second derivative will tell you the acceleration of a moving body. To find a higher order derivative, you just treat the first derivative as a new function and take its derivative in the ordinary way. You can keep doing this indefinitely.

29. For $y = x^4$, find the 1st through 6th derivatives.

Solve It

30. For $y = x^5 + 10x^3$, find the 1st, 2nd, 3rd, and 4th derivatives.

Solve It

31. For $y = \sin x + \cos x$, find the 1st through 6th derivatives.

Solve It

32. For $y = \cos x^2$, find the 1st, 2nd, and 3rd derivatives.

Solve It

Solutions for Differentiation Problems

1 $f(x) = 8$; $\boldsymbol{f'(x) = 0}$

The derivative of any constant is zero.

2 $g(x) = \pi^3$; $\boldsymbol{g'(x) = 0}$

Don't forget that even though π sort of looks like a variable (and even though other Greek letters like θ, α, and ω *are* variables), π is a number (roughly 3.14) and behaves like any other number. The same is true of $e \approx 2.718$. And when doing derivatives, constants like c and k also behave like ordinary numbers.

3 $g(x) = k\sin\frac{\pi}{2}\cos 2\pi$ (where k is a constant); $\boldsymbol{g'(x) = 0}$

If you feel bored because this first page of problems was so easy, just enjoy it; it won't last.

4 $f(x) = 5x^4$; $\boldsymbol{f'(x) = 20x^3}$

Bring the 4 in front and multiply it by the 5, and at the same time reduce the power by 1, from 4 to 3: $f'(x) = 20x^3$. Notice that the coefficient 5 has no effect on the derivative in the following sense: You can ignore the 5 temporarily, do the derivative of x^4 (which is $4x^3$), and then put the 5 back where it was and multiply it by 4.

5 $g(x) = \frac{-x^3}{10}$; $\boldsymbol{g'(x) = \frac{-3}{10}x^2}$

You can just write the derivative without any work: $\frac{-3x^2}{10}$. But if you want to do it more methodically, it works like this:

1. Rewrite $\frac{-x^3}{10}$ so you can see an ordinary coefficient: $-\frac{1}{10}x^3$.

2. Bring the 3 in front, multiply, and reduce the power by 1.

$$g'(x) = \frac{-3}{10}x^2 \text{ (which is the same, of course, as } \frac{-3x^2}{10}.)$$

6 $y = \sqrt{x^{-5}}$ $(x \geq 0)$; $\boldsymbol{y' = -\frac{5}{2}x^{-7/2}}$

Rewrite with an exponent and finish like problem 5 $\left(\sqrt{x^{-5}} = x^{-5/2}\right)$.

To write your answer without a negative power, you write $y' = -\frac{5}{2x^{7/2}}$ or $\frac{-5}{2x^{7/2}}$. Or you can write your answer without a fraction power, to wit: $y' = -\frac{5}{2\sqrt{x^7}}$ or $\frac{-5}{2\sqrt{x^7}}$ or $-\frac{5}{2\left(\sqrt{x}\right)^7}$ or $\frac{-5}{2\left(\sqrt{x}\right)^7}$. You say "po-*tay*-to," I say "po-*tah*-to."

7 $s(t) = 7t^6 + t + 10$; $\boldsymbol{s'(t) = 42t^5 + 1}$

Note that the derivative of plain old t or plain old x (or any other variable) is simply 1. In a sense, this is the simplest of all derivative rules, not counting the derivative of a constant. Yet for some reason, many people get it wrong. This is simply an example of the power rule: x is the same as x^1, so you bring the 1 in front and reduce the power by 1, from 1 to 0. That gives you $1x^0$. But because anything to the 0 power equals 1, you've got 1 times 1, which of course is 1.

8 $y = \left(x^3 - 6\right)^2$; $\boldsymbol{y' = 6x^5 - 36x^2}$

FOIL and then take the derivative.

$$= \left(x^3 - 6\right)\left(x^3 - 6\right)$$
$$= x^6 - 12x^3 + 36$$
$$y' = 6x^5 - 36x^2$$

9 $\frac{d}{dx}(x^3\cos x) = 3x^2\cos x - x^3\sin x$

TIP

Remember that $\frac{d}{dx}\cos x = -\sin x$. For a great mnemonic to remember the derivatives of trig functions, check out Chapter 15.

$$\frac{d}{dx}(x^3\cos x) = (x^3)'(\cos x) + (x^3)(\cos x)'$$
$$= 3x^2\cos x + x^3(-\sin x)$$
$$= 3x^2\cos x - x^3\sin x$$

10 $\frac{d}{dx}(\sin x\tan x) = \cos x\tan x + \sec x\tan x$

TIP

A helpful rule: $\frac{d}{dx}\tan x = \sec^2 x$.

$$\frac{d}{dx}(\sin x\tan x) = (\sin x)'(\tan x) + (\sin x)(\tan x)'$$
$$= \cos x\tan x + \sin x\sec^2 x$$
$$= \cos x\tan x + \sec x\tan x$$

11 $\frac{d}{dx}5x^3\ln x = 5x^2(3\ln x + 1)$

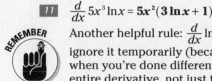

REMEMBER

Another helpful rule: $\frac{d}{dx}\ln x = \frac{1}{x}$. As for the 5, you can deal with it in two ways. First, you can ignore it temporarily (because coefficients have no effect on differentiation), and put it back when you're done differentiating. If you do it this way, don't forget that the "5" multiplies the entire derivative, not just the first term. The second way is probably easier and better: just make the "5" part of the first function. To wit:

$$\frac{d}{dx}(5x^3\ln x) = (5x^3)'(\ln x) + (5x^3)(\ln x)'$$
$$= 15x^2\ln x + 5x^3 \cdot \frac{1}{x}$$
$$= 15x^2\ln x + 5x^2 \quad \text{or}$$
$$= 5x^2(3\ln x + 1); \quad \text{take your pick.}$$

*12 $\frac{d}{dx}x^2 e^x\ln x = e^x\ln x(x^2 + 2x) + xe^x$

This is a challenge problem because, as you've probably noticed, there are three functions in this product instead of two. But it's a piece o' cake. Just make it two functions: either $(x^2 e^x)(\ln x)$ or $(x^2)(e^x\ln x)$. Take your pick.

REMEMBER

A handy rule: $\frac{d}{dx}e^x = e^x$ (Note that e^x and its multiples (like $4e^x$) are the only functions that are their own derivatives.)

1. Rewrite this "triple function" as the product of two functions.

$$= \frac{d}{dx}(x^2 e^x)(\ln x)$$

2. Apply the product rule.

$$\frac{d}{dx}(x^2 e^x)(\ln x) = (x^2 e^x)'(\ln x) + (x^2 e^x)(\ln x)'$$

3. Apply the product rule separately to $(x^2 e^x)'$, then substitute the answer back where it belongs.

$$(x^2 e^x)' = (x^2)'(e^x) + (x^2)(e^x)'$$
$$= 2xe^x + x^2 e^x$$

4. Complete the problem as shown in Step 2.

$$\left(x^2e^x\right)'(\ln x) + \left(x^2e^x\right)(\ln x)' = \left(2xe^x + x^2e^x\right)(\ln x) + x^2e^x \cdot \frac{1}{x}$$
$$= 2xe^x\ln x + x^2e^x\ln x + xe^x \ \text{ or}$$
$$= x^2e^x\ln x + 2xe^x\ln x + xe^x \ \text{ or}$$
$$= xe^x(x\ln x + 2\ln x + 1) \ \text{ or}$$
$$= e^x\ln x\left(x^2 + 2x\right) + xe^x$$

You say "pa-*ja*-mas," I say "pa-*jah*-mas."

13 $\dfrac{d}{dx}\dfrac{x^3}{\cos x} = \dfrac{3x^2\cos x + x^3\sin x}{\cos^2 x}$

$$\frac{d}{dx}\frac{x^3}{\cos x} = \frac{\left(x^3\right)'(\cos x) - \left(x^3\right)(\cos x)'}{(\cos x)^2}$$
$$= \frac{3x^2\cos x - x^3(-\sin x)}{\cos^2 x}$$
$$= \frac{3x^2\cos x + x^3\sin x}{\cos^2 x}$$

14 $\dfrac{d}{dx}\dfrac{\cos x}{e^x} = \dfrac{-\sin x - \cos x}{e^x}$

$$\frac{d}{dx}\frac{\cos x}{e^x} = \frac{(\cos x)'\left(e^x\right) - (\cos x)\left(e^x\right)'}{\left(e^x\right)^2}$$
$$= \frac{-e^x\sin x - e^x\cos x}{e^{2x}}$$
$$= \frac{-\sin x - \cos x}{e^x}$$

15 $\dfrac{d}{dx}\dfrac{3x^2+3}{\arctan x} = \dfrac{6x\arctan x - 3}{\arctan^2 x}$

A handy-dandy rule: $\dfrac{d}{dx}\arctan x = \dfrac{1}{1+x^2}$.

$$\frac{d}{dx}\frac{3x^2+3}{\arctan x} = \frac{\left(3x^2+3\right)'(\arctan x) - \left(3x^2+3\right)(\arctan x)'}{(\arctan x)^2}$$
$$= \frac{6x\arctan x - \left(3x^2+3\right)\left(\dfrac{1}{x^2+1}\right)}{\arctan^2 x}$$
$$= \frac{6x\arctan x - 3}{\arctan^2 x}$$

To remember the derivatives of the inverse trig functions, notice that the derivative of each co-function (*arccosine, arccotangent,* and *arccosecant*) is the negative of its corresponding function. So, you really only need to memorize the derivatives of *arcsin, arctan,* and *arcsec.* These three have a 1 in the numerator. The two that contain the letter "s," *arcsin* and *arcsec,* contain a square root in the denominator. *Arctan* has no "s" and no square root.

* 16 $\dfrac{d}{dx}\dfrac{\sin x}{x^3\ln x} = \dfrac{x\cos x\ln x - 3\sin x\ln x - \sin x}{x^4(\ln x)^2}$

$$\dfrac{d}{dx}\dfrac{\sin x}{x^3\ln x} = \dfrac{(\sin x)'(x^3\ln x) - (\sin x)(x^3\ln x)'}{(x^3\ln x)^2}$$

$$= \dfrac{(\cos x)(x^3\ln x) - (\sin x)\overbrace{\left((x^3)'(\ln x) + (x^3)(\ln x)'\right)}^{\text{Product Rule}}}{x^6(\ln x)^2}$$

$$= \dfrac{x^3\cos x\ln x - (\sin x)\left(3x^2\ln x + (x^3)\left(\dfrac{1}{x}\right)\right)}{x^6(\ln x)^2}$$

$$= \dfrac{x^3\cos x\ln x - 3x^2\sin x\ln x - x^2\sin x}{x^6(\ln x)^2}$$

$$= \dfrac{x\cos x\ln x - 3\sin x\ln x - \sin x}{x^4(\ln x)^2}$$

17 $f(x) = \sin x^2;$ $f'(x) = 2x\cos x^2$

Because the argument of the sine function is something other than a plain old x, this is a chain rule problem. Just use the rule for the derivative of sine, not touching the inside stuff (x^2), then multiply your result by the derivative of x^2.

$$f'(x) = \cos(x^2)\cdot 2x$$
$$= 2x\cos x^2$$

18 $g(x) = \sin^3 x;$ $g'(x) = 3\sin^2 x\cos x$

Rewrite $\sin^3 x$ as $(\sin x)^3$ so that it's clear that the outermost function is the cubing function. By the chain rule, the derivative of *stuff*^{3} is $3\,stuff^2\cdot stuff'$. The *stuff* here is $\sin x$ and thus *stuff'* is $\cos x$. So your final answer is $3(\sin x)^2\cdot\cos x$, or $3\sin^2 x\cos x$.

19 $s(t) = \tan(\ln t);$ $s'(t) = \sec^2(\ln t)\cdot\dfrac{1}{t}$

The derivative of tan is $\sec^2$, so the derivative of tan(*lump*) is $\sec^2(lump)\cdot lump'$. You better know by now that the derivative of $\ln t$ is $\dfrac{1}{t}$, so your final result is $\sec^2(\ln t)\cdot\dfrac{1}{t}$.

20 $y = e^{4x^3};$ $y' = 12x^2 e^{4x^3}$

The derivative of e^x is e^x, so by the chain rule, the derivative of e^{glob} is $e^{glob}\cdot glob'$. So $y' = e^{4x^3}\cdot 12x^2$, or $12x^2 e^{4x^3}$.

* 21 $f(x) = x^4\sin^3 x;$ $f'(x) = 4x^3\sin^3 x + 3x^4\sin^2 x\cos x$

This problem involves both the product rule and the chain rule. Which do you do first? Note that the chain rule part of this problem, $\sin^3 x$, is one of the two things being multiplied, so it is part of — or sort of *inside* — the product. And, like with pure chain rule problems, with problems involving more than one rule, you work from outside, in. So here you begin with the product rule. Here's another way to look at it:

TIP   If you're not sure about the order of the rules in a complicated derivative problem, imagine that you plugged a number into x in the original function and had to compute the answer. Your *last* computation tells you where to start. If, for example, you plugged 2 into $x^4\sin^3 x$, you would compute 2^4, then $\sin 2$, then you'd cube that to get $\sin^3 2$, and, finally, you'd multiply 2^4 by $\sin^3 2$. Because your final step was *multiplication*, you begin with the *product* rule.

$$f(x) = x^4 \sin^3 x$$
$$= (x^4)(\sin^3 x)$$
$$f'(x) = (x^4)'(\sin^3 x) + (x^4)(\sin^3 x)' \quad \text{(product rule)}$$

Use the chain rule to solve $(\sin^3 x)'$, then go back and finish the problem. $\sin^3 x$ means $(\sin x)^3$ and that's *stuff*^{3}. The derivative of *stuff*^{3} is $3\,\textit{stuff}^2 \cdot \textit{stuff}'$, so the derivative of $(\sin x)^3$ is $3(\sin x)^2 \cdot \cos x$. Now continue the solution.

$$f'(x) = (x^4)'(\sin^3 x) + (x^4)(\sin^3 x)'$$
$$= 4x^3 \sin^3 x + x^4 \cdot 3(\sin x)^2 \cos x$$
$$= 4x^3 \sin^3 x + 3x^4 \sin^2 x \cos x$$

*22 $\quad g(x) = \dfrac{(\ln x)^2}{5x-4}; \quad g'(x) = \dfrac{2\ln x}{x(5x-4)} - \dfrac{5(\ln x)^2}{(5x-4)^2}$

Here you've got the chain rule inside the quotient rule. Start with the quotient rule:

$$g'(x) = \frac{\left((\ln x)^2\right)'(5x-4) - (\ln x)^2(5x-4)'}{(5x-4)^2}$$

Next, take care of the chain rule solution for $\left((\ln x)^2\right)'$. You want the derivative of *glob*^{2} — that's $2\,\textit{glob} \cdot \textit{glob}'$. So the derivative of $(\ln x)^2$ is $2(\ln x)\left(\frac{1}{x}\right)$. Now you can finish:

$$g'(x) = \frac{2(\ln x)\left(\frac{1}{x}\right)(5x-4) - (\ln x)^2(5)}{(5x-4)^2}$$
$$= \frac{(10x-8)\ln x - 5x(\ln x)^2}{x(5x-4)^2}$$
$$= \frac{2\ln x}{x(5x-4)} - \frac{5(\ln x)^2}{(5x-4)^2}$$

*23 $\quad y = \cos^3 4x^2; \quad y' = -24x\cos^2 4x^2 \sin 4x^2$

Triply nested!

$$y = \left(\cos 4x^2\right)^3$$

The derivative of *stuff*^{3} is $3\,\textit{stuff}^2 \cdot \textit{stuff}'$, so, you've got,

$$y' = 3\left(\cos 4x^2\right)^2 \cdot \left(\cos 4x^2\right)'$$

Now you do the derivative of $\cos(\textit{glob})$, which is $-\sin(\textit{glob}) \cdot \textit{glob}'$. Two down, one to go:

$$y' = 3\left(\cos 4x^2\right)^2(-\sin 4x^2) \cdot (4x^2)'$$
$$= 3\cos^2 4x^2(-\sin 4x^2) \cdot 8x$$
$$= -24x\cos^2 4x^2 \sin 4x^2$$

* **24** $\dfrac{d}{dx}\tan^3 e^{x^2} = \mathbf{6}\mathbf{x}\mathbf{e}^{x^2}\tan^2 e^{x^2}\sec^2 e^{x^2}$

Holy quadrupely nested quadruple nestedness, Batman! This is one for the Riddler.

$$= \frac{d}{dx}\left(\tan e^{x^2}\right)^3$$

$$= 3\left(\tan e^{x^2}\right)^2 \cdot \left(\tan e^{x^2}\right)' \qquad \left(because\ \frac{d}{dx}\ stuff^3 = 3stuff^2 \cdot stuff'\right)$$

$$= 3\tan^2 e^{x^2}\sec^2 e^{x^2}\cdot\left(e^{x^2}\right)' \qquad \left(because\ \frac{d}{dx}\tan(glob) = \sec^2(glob)\cdot glob'\right)$$

$$= 3\tan^2 e^{x^2}\sec^2 e^{x^2}\cdot e^{x^2}\cdot\left(x^2\right)' \qquad \left(because\ \frac{d}{dx}e^{lump} = e^{lump}\cdot lump'\right)$$

$$= 3\tan^2 e^{x^2}\sec^2 e^{x^2}\cdot e^{x^2}\cdot 2x$$

$$= 6xe^{x^2}\tan^2 e^{x^2}\sec^2 e^{x^2}$$

POW!!   CRUNCH!   BAM!!!

25 If $y^3 - x^2 = x + y,$ $\ y' = \dfrac{2x+1}{3y^2-1}$

1. Take the derivative of all four terms, using the chain rule (sort of) for all terms containing a y.

$$3y^2\,y' - 2x = 1 + y'$$

2. Move all terms containing y' to the left, all other terms to the right, and factor out y'.

$$3y^2\,y' - y' = 1 + 2x$$
$$y'\left(3y^2 - 1\right) = 1 + 2x$$

3. Divide and voilà!

$$y' = \frac{2x+1}{3y^2-1}$$

26 If $3y + \ln y = 4e^x,$ $\ y' = \dfrac{4ye^x}{3y+1}$

Follow steps for problem 25.

$$3y' + \frac{1}{y}\,y' = 4e^x$$

$$y'\left(3 + \frac{1}{y}\right) = 4e^x$$

$$y' = \frac{4e^x}{3 + \dfrac{1}{y}}$$

$$= \frac{4ye^x}{3y+1}$$

27 For $x^2 y = y^3 x + 5y + x$, find $\dfrac{dy}{dx}$ by implicit differentiation. $y' = \dfrac{y^3 - 2xy + 1}{-3y^2 x + x^2 - 5}$

This time you've got two *products* to deal with, so use the product rule for the two products and the regular rules for the other two terms.

$$(x^2)' y + x^2 y' = (y^3)' x + y^3 x' + 5y' + 1$$
$$2xy + x^2 y' = 3y^2 y' x + y^3 + 5y' + 1$$
$$x^2 y' - 3y^2 y' x - 5y' = y^3 + 1 - 2xy$$
$$y'(x^2 - 3y^2 x - 5) = y^3 - 2xy + 1$$
$$y' = \frac{y^3 - 2xy + 1}{-3y^2 x + x^2 - 5}$$

* 28 If $y + \cos^2 y^3 = \sin 5x^2$, find the slope of the curve at $\left(\sqrt{\dfrac{\pi}{10}}, 0 \right)$. **The slope is zero.**

You need a slope, so you need the derivative.

$$\underbrace{y'}_{\substack{\text{Implicit} \\ \text{Differentiation}}} + \underbrace{2\cos y^3 \cdot (-\sin y^3)(y^3)'}_{\substack{\text{Chain Rule} \\ (\text{twice nested})}} = \underbrace{\cos(5x^2)(10x)}_{\text{Chain Rule}}$$

$$y' + 2\cos y^3 (-\sin y^3)\underbrace{(3y^2 y')}_{\substack{\text{Implicit} \\ \text{Differentiation}}} = 10x \cos 5x^2$$

$$y'(1 - 6y^2 \cos y^3 \sin y^3) = 10x \cos 5x^2$$

$$y' = \frac{10x \cos 5x^2}{1 - 6y^2 \cos y^3 \sin y^3}$$

You need the slope at $x = \sqrt{\dfrac{\pi}{10}}$, $y = 0$, so plug those numbers in to the derivative. Actually, you can save yourself a lot of work if you notice that the numerator will equal zero (because $\cos\left(5\sqrt{\dfrac{\pi}{10}}^2 \right) = 0$) and the denominator will equal 1 (because $y = 0$). And thus the slope of the curve at this point is zero. A tangent line with a zero slope is horizontal, and because this tangent line touches the curve where $y = 0$, the tangent line is the x-axis.

29 For $y = x^4$, find the 1st through 6th derivatives.

$$y' = 4x^3$$
$$y'' = 12x^2$$
$$y''' = 24x$$
$$y^{(4)} = 24$$
$$y^{(5)} = 0$$
$$y^{(6)} = 0$$

Extra credit: $y^{(2005)} = 0$

30 For $y = x^5 + 10x^3$ find the 1st, 2nd, 3rd, and 4th derivatives.

$$y' = 5x^4 + 30x^2$$
$$y'' = 20x^3 + 60x$$
$$y''' = 60x^2 + 60$$
$$y^{(4)} = 120x$$

31 For $y = \sin x + \cos x$, find the 1st through 6th derivatives.

$$y' = \cos x - \sin x$$
$$y'' = -\sin x - \cos x$$
$$y''' = -\cos x + \sin x$$
$$y^{(4)} = \sin x + \cos x$$
$$y^{(5)} = \cos x - \sin x$$
$$y^{(6)} = -\sin x - \cos x$$

Notice that the 4th derivative equals the original function; that the 5th derivative equals the 1st, and so on. This cycle of four functions repeats ad infinitum.

32 For $y = \cos x^2$, find the 1st, 2nd, and 3rd derivatives.

$$y = \cos x^2$$
$$y' = -2x \sin x^2 \quad \textbf{(chain rule)}$$
$$y'' = (-2x)'(\sin x^2) + (-2x)(\sin x^2)' \quad \text{(product rule)}$$
$$= -2\sin x^2 - 2x\cos(x^2)2x \quad \text{(chain rule)}$$
$$= -2\sin x^2 - 4x^2 \cos x^2$$
$$y''' = -2\cos(x^2)2x - \left[(4x^2)'(\cos x^2) + (4x^2)(\cos x^2)'\right]$$
$$= -4x\cos x^2 - \left[8x\cos x^2 + 4x^2(-\sin(x^2)2x)\right]$$
$$= -4x\cos x^2 - 8x\cos x^2 + 8x^3 \sin x^2$$
$$= 8x^3 \sin x^2 - 12x \cos x^2$$

Chapter 7

Analyzing Those Shapely Curves with the Derivative

In This Chapter

▶ Mum's the word: Minimum, maximum, extremum

▶ Concavity and inflection points

▶ The nasty mean value theorem

This chapter gives you lots of practice using the derivative to analyze the shape of curves and their significant features and points. Don't forget: The derivative tells you the slope of a curve, so any problem involving anything about the slope or steepness of a curve is a derivative problem.

The First Derivative Test and Local Extrema

One of the most common applications of the derivative employs the simple idea that at the top of a hill or at the bottom of a valley, you're neither going up nor down; in other words, there's no steepness and the slope — and thus the derivative — equals zero. You can therefore use the derivative to locate the top of "hills" and the bottom of "valleys," called *local extrema*, on just about any function by setting the derivative of the function equal to zero and solving.

Q. Use the first derivative test to determine the location of the local extrema of $g(x) = 15x^3 - x^5$. See the following figure.

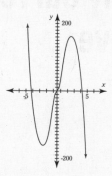

A. The local min is at $(-3, -162)$, and the local max is at $(3, 162)$.

1. Find the first derivative of g using the power rule.

$$g(x) = 15x^3 - x^5$$
$$g'(x) = 45x^2 - 5x^4$$

2. Set the derivative equal to zero and solve for x to get the critical numbers of g.

$$45x^2 - 5x^4 = 0$$
$$5x^2(9 - x^2) = 0$$
$$5x^2(3 - x)(3 + x) = 0$$

$$5x^2 = 0 \text{ or } 3 - x = 0 \text{ or } 3 + x = 0$$
$$x = 0, 3, \text{ or } -3$$

If the first derivative were undefined for some x-values in the domain of g, there could be more critical numbers, but because $g'(x) = 45x^2 - 5x^4$ is defined for all real numbers, $0, 3, -3$ is the complete list of critical numbers of g.

Remember: If f is defined at a number c and the derivative at $x = c$ is either zero or undefined, then c is a critical number of f.

3. Plot the three critical numbers on a number line, noting that they create four regions (see the figure in Step 5).

4. Test a number from each of the four regions, noting whether the results are positive or negative.

Note that if you use round numbers like 0, –10, or 10, you can often do the arithmetic in your head.

$$g'(-10) = 45(-10)^2 - 5(-10)^4 = -45,500$$
$$g'(-1) = 45(-1)^2 - 5(-1)^4 = 40$$
$$g'(1) = 45(1)^2 - 5(1)^4 = 40$$
$$g'(10) = 45(10)^2 - 5(10)^4 = -45,500$$

5. Draw a "sign graph." Take your number line and label each region — based on your results from Step 4 — positive (increasing) or negative (decreasing). See the following figure.

decreasing   increasing   increasing   decreasing
   −   +   +   −

        -3   0   3

This sign graph tells you where the function is rising or increasing and where it is falling or decreasing.

6. Use the sign graph to determine whether there's a local minimum, local maximum, or neither at each critical number.

Because g goes down on its way to $x = -3$ and up after $x = -3$, it must bottom out at $x = -3$, so there's a local min there. Conversely, g peaks at $x = 3$ because it rises until $x = 3$, then falls. There is thus a local max at $x = 3$. And because g climbs on its way to $x = 0$ and then climbs further, there is neither a min nor a max at $x = 0$.

7. Determine the y-values of the local extrema by plugging the x-values into the original function.

$$g(-3) = 15(-3)^3 - (-3)^5$$
$$= -162$$

$$g(3) = 15(3)^3 - (3)^5$$
$$= 162$$

So the local min is at $(-3, -162)$, and the local max is at $(3, 162)$.

1. Use the first derivative to find the local extrema of $f(x) = 6x^{2/3} - 4x + 1$. *Tip*: You better write small if you want to do this problem on half a page.

Solve It

2. Find the local extrema of $h(x) = \frac{x}{\sqrt{2}} + \cos x - \frac{\sqrt{2}}{2}$ in the interval $(0, 2\pi)$ with the first derivative test.

Solve It

3. Locate the local extrema of $y = \left(x^2 - 8\right)^{2/3}$ with the first derivative test.

Solve It

4. Using the first derivative test, determine the local extrema of $s = \dfrac{t^4 + 4}{-2t^2}$.

Solve It

The Second Derivative Test and Local Extrema

With the second derivative test, you use — can you guess? — the *second* derivative to test for local extrema. The second derivative test is based on the absolutely brilliant idea that the crest of a hill has a hump shape (∩) and the bottom of a valley has a trough shape (∪).

After you find a function's critical numbers, you've got to decide whether to use the first or the second derivative test to find the extrema. For some functions, the second derivative test is the easier of the two because 1) The second derivative is usually easy to get, 2) You can often plug the critical numbers into the second derivative and do a quick computation, and 3) you will often get non-zero results and thus get your answers without having to do a sign graph and test regions. On the other hand, testing regions on a sign graph (the first derivative test) is also fairly quick and easy, and if the second derivative test fails (see the warning), you'll have to do that anyway. As you do practice problems, you'll get a feel for when to use each test.

When the second derivative equals zero, the second derivative test fails and you learn nothing about local extrema. You've then got to use the first derivative test to determine whether there's a local *extremum* (the singular of extrema) there.

Q. Take the function from the example in the previous section, $g(x) = 15x^3 - x^5$, but this time find its local extrema using the second derivative test.

A. **The local min is at $x = -3$ and the local max is at $x = 3$.**

You've probably figured out that to use the second derivative test you need the second derivative:

$$g(x) = 15x^3 - x^5$$
$$g'(x) = 45x^2 - 5x^4$$
$$g''(x) = 90x - 20x^3$$

Now all you do is plug in the critical numbers of g from the example in the last section:

$$g''(-3) = 270$$
$$g''(0) = 0$$
$$g''(3) = -270$$

The fact that $g''(-3)$ is *positive* tells you that g is concave *up* (∪) there, and thus that there's a local *min*. And the fact that $g''(3)$ is *negative* tells you that g is concave *down* (∩) at $x = 3$, and, therefore, that there's a local *max* there. And, while it may seem that $g''(0) = 0$ confirms what you figured out previously (that there's neither a min nor a max there), you actually learn nothing when the second derivative is zero; you have to use the first derivative test.

Tip: If you only have one critical point between a local min and a local max (and no discontinuities), it has to be an inflection point, and if you have a single critical number between two known maxes (see problem 7), the only possibility for the middle critical number is a local min (and vice-versa). So in these cases, it really doesn't matter if the second derivative test fails with the middle critical number. If this not-by-the-book reasoning doesn't work for your calc teacher, you might say (with just a touch of sarcasm in your voice), "Oh, so in other words, you've got something against logic and common sense."

5. Use the second derivative test to analyze the critical numbers of the function from problem 2, $h(x) = \dfrac{x}{\sqrt{2}} + \cos x - \dfrac{\sqrt{2}}{2}$.

Solve It

6. Find the local extrema of $f(x) = -2x^3 + 6x^2 + 1$ with the second derivative test.

Solve It

7. Find the local extrema of $y = 2x^4 - \frac{1}{3}x^6$ with the second derivative test.

Solve It

8. Consider the function from problem 3, $y = (x^2 - 8)^{2/3}$, and the function $s = 8 + \frac{21t}{4} - \frac{7t^3}{4}$. Which of the two functions is easier to analyze with the second derivative test, and why? For the function you pick, use the second derivative test to find its local extrema.

Solve It

Finding Mount Everest: Absolute Extrema

The basic idea in this section is quite simple. Instead of finding all local extrema like in the previous sections (all the peaks and all the valleys), you just want to determine the single highest point and single lowest point along a *continuous* function in some *closed* interval. These *absolute extrema* can occur at a peak or valley *or* at an edge(s) of the interval. (Note: You could have, say, two peaks at the same height so there'd be a tie for the absolute max; but there would still be exactly one *y-value* that's the absolute maximum value on the interval.)

Before you practice with some problems, look at Figure 7-1 to see two standard absolute extrema problems (*continuous* functions on a *closed* interval) and at Figure 7-2 for four strange functions that don't have the standard single absolute max and single absolute min.

Figure 7-1:
Two
standard
absolute
extrema
functions.

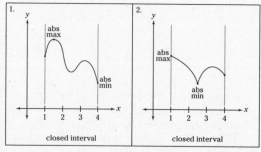

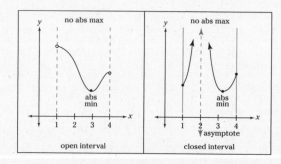

Figure 7-2:
Four non-
standard
absolute
extrema
functions.

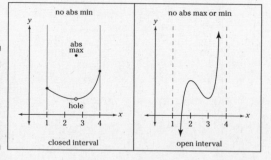

Q. Determine the absolute min and absolute max of $f(x) = \sqrt{|x|} - x$ in the interval $[-1, \frac{1}{2}]$.

A. **The absolute max is 2 and the absolute min is 0.**

1. Get all the critical numbers:

First, set the derivative equal to zero (note, first split this function in two to get rid of the absolute value bars):

$$f(x) = \sqrt{x} - x \quad (x \geq 0)$$

$$f'(x) = \frac{1}{2\sqrt{x}} - 1$$

$$0 = \frac{1}{2\sqrt{x}} - 1$$

$$2\sqrt{x} = 1$$

$$x = \frac{1}{4}$$

$$f(x) = \sqrt{-x} - x \quad (x < 0)$$

$$f'(x) = \frac{-1}{2\sqrt{-x}} - 1$$

$$0 = \frac{-1}{2\sqrt{-x}} - 1$$

$$2\sqrt{-x} = -1$$

No solution

Now, determine whether the derivative is undefined anywhere:

The derivative is undefined at $x = 0$ because the denominator of the derivative can't equal zero. (If you graph this function (always a good idea) you'll also see the cusp at $x = 0$ and thus know immediately that there's no derivative there.)

The critical numbers are therefore 0 and $\frac{1}{4}$.

2. Compute the function values (the heights) at all the critical numbers.

$$f\left(\frac{1}{4}\right) = \frac{1}{4} \qquad f(0) = 0$$

It's just a coincidence, by the way, that in both cases the argument equals the answer.

3. Compute the function values at the two edges of the interval.

$$f(-1) = 2 \qquad f\left(\frac{1}{2}\right) = \left(\frac{\sqrt{2}-1}{2}\right) \approx 0.207$$

4. The highest of all the function values from Steps 2 and 3 is the absolute max; the lowest of all the values from Steps 2 and 3 is the absolute min.

Thus, 2 is the absolute max and 0 is the absolute min.

Note that finding absolute extrema involves less work than finding local extrema because you don't have to use the first or second derivative tests — do you see why?

9. Find the absolute extrema of
$f(x) = \sin x + \cos x$ on the interval $[0, 2\pi]$.

Solve It

10. Find the absolute extrema of
$g(x) = 2x^3 - 3x^2 - 5$ on the interval $[-.5, .5]$.

Solve It

11. Find the absolute extrema of
$p(x) = (x+1)^{4/5} - .5x$ on the interval
$[-2, 31]$.

Solve It

12. Find the absolute extrema of
$q(x) = 2\cos 2x + 4\sin x$ on the interval
$\left[-\dfrac{\pi}{2}, \dfrac{5\pi}{4}\right]$.

Solve It

Smiles and Frowns: Concavity and Inflection Points

Another purpose of the second derivative is to analyze concavity and points of inflection. (For a refresher, look at Figure 7-3: The area between A and B is concave down — like an upside-down spoon or a frown; the areas on the outsides of A and B are concave up — right-side up spoon or a smile; and A and B are inflection points.) A positive second derivative means concave up; a negative second derivative means concave down. Where the concavity switches from up to down or down to up, you've got an inflection point, and the second derivative there will be zero (or sometimes undefined).

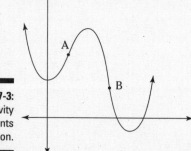

Figure 7-3:
Concavity
and points
of inflection.

All inflection points have a second derivative of zero (if the second derivative exists), but not all points with a second derivative of zero are inflection points. This is no different from "all ships are boats but not all boats are ships."

However, you can have an inflection point where the second derivative is undefined. This occurs when the inflection point has a vertical tangent and in some bizarre curves that you shouldn't worry about that have a weird discontinuity in the second derivative.

Q. Find the intervals of concavity and the inflection points of $f(x) = 3x^5 - 5x^3 + 10$. Note that the following solution is analogous to the solution for finding local extrema with the first derivative.

A. *f* is concave down from $-\infty$ to the inflection point at $\left(-\dfrac{\sqrt{2}}{2}, \sim 11.24\right)$; concave up from there to the inflection point at (0, 10); concave down from there to the third inflection point at $\left(\dfrac{\sqrt{2}}{2}, \sim 8.76\right)$; and concave up from there to ∞.

1. Find the second derivative of *f*.

$$f(x) = 3x^5 - 5x^3 + 10$$
$$f'(x) = 15x^4 - 15x^2$$
$$f''(x) = 60x^3 - 30x$$

2. Set the second derivative equal to zero and solve.

$$60x^3 - 30x = 0$$
$$30x(2x^2 - 1) = 0$$

$$2x^2 - 1 = 0$$
$$2x^2 = 1$$
$$30x = 0$$
$$x = 0 \quad \text{or} \quad x^2 = \frac{1}{2}$$
$$x = \pm\frac{\sqrt{2}}{2}$$

3. Check whether there are any *x*-values where the second derivative is undefined.

There are none, so $-\dfrac{\sqrt{2}}{2}$, 0, and $\dfrac{\sqrt{2}}{2}$ are the three second derivative "critical numbers." (Technically these aren't called critical numbers, but they could be because they work just like first derivative critical numbers.)

4. Plot these "critical numbers" on a number line and test the regions.

You can use –1, –.5, .5, and 1 as test numbers. The following figure shows the second derivative sign graph.

$$f''(x) = 60x^3 - 30x$$

$$f''(-1) = -30$$
$$f''(-.5) = 7.5$$
$$f''(.5) = -7.5$$
$$f''(1) = 30$$

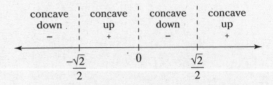

Because the concavity switches at all these "critical numbers" and because the second derivative exists at those numbers (from Steps 2 and 3), there are inflection points at those three *x*-values. (If the concavity switches at a point where the second derivative is undefined, you have to check one more thing before concluding that you've got an inflection point: whether you can draw a tangent line there. This is the case when the first derivative is defined or if there's a vertical tangent.) In a nutshell, if the concavity switches at a point where the curve is smooth, you've got an inflection point.

5. Determine the location of the three inflection points.

$$f(x) = 3x^5 - 5x^3 + 10$$
$$f\left(-\frac{\sqrt{2}}{2}\right) \approx 11.24$$
$$f(0) = 10$$
$$f\left(\frac{\sqrt{2}}{2}\right) \approx 8.76$$

So *f* is concave down from $-\infty$ to the inflection point at $\left(-\dfrac{\sqrt{2}}{2}, \sim 11.24\right)$, concave up from there to the inflection point at (0, 10), concave down from there to the third inflection point at $\left(\dfrac{\sqrt{2}}{2}, \sim 8.76\right)$, and, finally, concave up from there to ∞.

13. Find the intervals of concavity and the inflection points of $f(x) = -2x^3 + 6x^2 - 10x + 5$.

Solve It

14. Find the intervals of concavity and the inflection points of $g(x) = x^4 - 12x^2$.

Solve It

15. Find the intervals of concavity and the inflection points of $p(x) = \dfrac{x}{x^2 + 9}$.

Solve It

16. Find the intervals of concavity and the inflection points of $q(x) = \sqrt[5]{x} - \sqrt[3]{x}$.

Solve It

The Mean Value Theorem: Go Ahead, Make My Day

The *Mean Value Theorem* is based on an incredibly simple idea. Say you go for a one hour drive and travel 50 miles. Your average speed, of course, would be 50 mph. The Mean Value Theorem says that there must be at least one point during your trip when your speed was exactly 50 mph. But you don't need a fancy-pants calculus theorem to tell you that. It's just common sense. If you went slower than 50 mph the whole way, you couldn't average 50. And if you went faster than 50 the whole way (this assumes you're going faster than 50 at your starting point), your average speed would be greater than 50. The only way to average 50 is to go exactly 50 the whole way or to go slower than 50 some times and faster than 50 at other times. And when you speed up or slow down, you have to pass exactly 50 at some point.

With the Mean Value Theorem, you figure an average rate or slope over an interval and then use the first derivative to find one or more points in the interval where the instantaneous rate or slope equals the average rate or slope. Here's an example:

Q. Given $f(x) = x^3 - 4x^2 - 5x$, find all numbers c in the open interval $(2, 4)$ where the instantaneous rate equals the average rate over the interval.

A. The only answer is $\dfrac{4 + 2\sqrt{7}}{3}$.

Basically, you're finding the points along the curve in the interval where the slope is the same as the slope from $\left(2, f(2)\right)$ to $\left(4, f(4)\right)$. Mathematically speaking, find all numbers c where

$$f'(c) = \frac{f(4) - f(2)}{4 - 2}.$$

1. Get the first derivative.

$$f(x) = x^3 - 4x^2 - 5x$$
$$f'(x) = 3x^2 - 8x - 5$$

2. Using the slope formula, $m = \dfrac{y_2 - y_1}{x_2 - x_1}$, figure the slope from $\left(2, f(2)\right)$ to $\left(4, f(4)\right)$.

$$f(4) = 4^3 - 4 \cdot 4^2 - 5 \cdot 4$$
$$= -20$$

$$f(2) = 2^3 - 4 \cdot 2^2 - 5 \cdot 2$$
$$= -18$$

$$m = \frac{f(4) - f(2)}{4 - 2}$$
$$= \frac{-20 - (-18)}{2}$$
$$= -1$$

3. Set the derivative equal to this slope and solve.

$$3x^2 - 8x - 5 = -1$$
$$3x^2 - 8x - 4 = 0$$

$$x = \frac{8 \pm \sqrt{(-8)^2 - 4(3)(-4)}}{6}$$

$$= \frac{8 \pm 4\sqrt{7}}{6}$$

$$= \frac{4 + 2\sqrt{7}}{3} \text{ or } \frac{4 - 2\sqrt{7}}{3}$$

$$\approx 3.10 \text{ or } \approx -.43$$

Because $-.43$ is outside the interval $(2, 4)$, your only answer is $\dfrac{4 + 2\sqrt{7}}{3}$.

By the way, the Mean Value Theorem only works for functions that are differentiable over the open interval in question and continuous over the open interval and its endpoints.

17. For $g(x) = x^3 + x^2 - x$, find all the values c in the interval $(-2, 1)$ that satisfy the Mean Value Theorem.

Solve It

18. For $s(t) = t^{4/3} - 3t^{1/3}$, find all the values c in the interval $(0, 3)$ that satisfy the Mean Value Theorem.

Solve It

Solutions for Derivatives and Shapes of Curves

_1_ Use the first derivative to find the local extrema of $f(x) = 6x^{2/3} - 4x + 1$. **Local min at (0, 1); local max at (1, 3).**

1. Find the first derivative using the power rule.

$$f(x) = 6x^{2/3} - 4x + 1$$
$$f'(x) = 4x^{-1/3} - 4$$

2. Find the critical numbers of f.

   a. Set the derivative equal to zero and solve.

   $$4x^{-1/3} - 4 = 0$$
   $$x^{-1/3} = 1$$
   $$x = 1$$

   b. Determine the x-values where the derivative is undefined: $f'(x) = 4x^{-1/3} - 4 = \dfrac{4}{\sqrt[3]{x}} - 4$

   Because the denominator is not allowed to equal zero, $f'(x)$ is undefined at $x = 0$. Thus the critical numbers of f are 0 and 1.

3. Plot the critical numbers on a number line.

   I'm going to skip the figure this time because I assume you can imagine a number line with dots at 0 and 1. Don't disappoint me!

4. Plug a number from each of the three regions into the derivative.

$$f'(-1) = 4(-1)^{-1/3} - 4 = -4 - 4 \quad = -8$$
$$f'\left(\frac{1}{2}\right) = 4\left(\frac{1}{2}\right)^{-1/3} - 4 = 4(2)^{1/3} - 4 = positive$$
$$f'(8) = 4(8)^{-1/3} - 4 = 2 - 4 \quad = -2$$

 Note, again, how the numbers I picked for the first and third computations made the math easy. With the second computation, you can save a little time and skip the final calculation because all you care about is whether the result is positive or negative — this assumes that you know that the cube root of 2 is more than 1 (you better!).

5. Draw your sign graph (see the following figure).

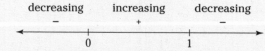

6. Determine whether there's a local min or max or neither at each critical number.

   f goes down to $x = 0$ and then up, so there's a local min at $x = 0$, and f goes up to $x = 1$ and then down, so there's a local max at $x = 1$.

7. Figure the y-value of the two local extrema.

$$f(0) = 6(0)^{2/3} - 4(0) + 1 = 1$$
$$f(1) = 6(1)^{2/3} - 4(1) + 1 = 3$$

Thus, there's a local min at (0, 1) and a local max at (1, 3). Check this answer by looking at a graph of f on your graphing calculator.

2 Find the local extrema of $h(x) = \frac{x}{\sqrt{2}} + \cos x - \frac{\sqrt{2}}{2}$ in the interval $(0, 2\pi)$ with the first derivative test. **Local max at** $\left(\frac{\pi}{4}, \frac{\pi\sqrt{2}}{8}\right)$; **local min at** $\left(\frac{3\pi}{4}, \frac{3\pi\sqrt{2}}{8} - \sqrt{2}\right)$.

1. Find the first derivative.

$$h(x) = \frac{x}{\sqrt{2}} + \cos x - \frac{\sqrt{2}}{2}$$

$$h'(x) = \frac{1}{\sqrt{2}} - \sin x$$

2. Find the critical numbers of h.

   a. Set the derivative equal to zero and solve:

$$\frac{1}{\sqrt{2}} - \sin x = 0$$

$$\sin x = \frac{\sqrt{2}}{2}$$

$$x = \frac{\pi}{4} \text{ or } \frac{3\pi}{4} \quad \text{(These are the solutions in the given interval.)}$$

   b. Determine the x-values where the derivative is undefined.

   The derivative isn't undefined anywhere, so the critical numbers of h are $\frac{\pi}{4}$ and $\frac{3\pi}{4}$.

3. Test numbers from each region on your number line.

$$h'\left(\frac{\pi}{6}\right) = \frac{1}{\sqrt{2}} - \sin\frac{\pi}{6} \qquad h'\left(\frac{\pi}{2}\right) = \frac{\sqrt{2}}{2} - \sin\frac{\pi}{2} \qquad h'(\pi) = \frac{\sqrt{2}}{2} - \sin\pi$$

$$= \frac{\sqrt{2}}{2} - \frac{1}{2} \qquad\qquad = \frac{\sqrt{2}}{2} - 1 \qquad\qquad = \frac{\sqrt{2}}{2} - 0$$

$$= positive \qquad\qquad = negative \qquad\qquad = positive$$

4. Draw a sign graph (see the following figure).

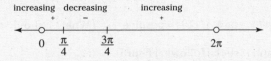

5. Decide whether there's a local min, max, or neither at each of the two critical numbers.

   Going from left to right along the function, you go up until $x = \frac{\pi}{4}$ and then down, so there's a local max at $x = \frac{\pi}{4}$. It's vice-versa for $x = \frac{3\pi}{4}$, so there's a local min there.

6. Compute the y-values of these two extrema.

$$h\left(\frac{\pi}{4}\right) = \frac{\frac{\pi}{4}}{\sqrt{2}} + \cos\frac{\pi}{4} - \frac{\sqrt{2}}{2} \qquad\qquad h\left(\frac{3\pi}{4}\right) = \frac{\frac{3\pi}{4}}{\sqrt{2}} + \cos\frac{3\pi}{4} - \frac{\sqrt{2}}{2}$$

$$= \frac{\pi}{4\sqrt{2}} + \frac{\sqrt{2}}{2} - \frac{\sqrt{2}}{2} \qquad\qquad = \frac{3\pi\sqrt{2}}{8} - \frac{\sqrt{2}}{2} - \frac{\sqrt{2}}{2}$$

$$= \frac{\pi\sqrt{2}}{8} \qquad\qquad\qquad = \frac{3\pi\sqrt{2}}{8} - \sqrt{2}$$

So you've got a max at $\left(\frac{\pi}{4}, \frac{\pi\sqrt{2}}{8}\right)$ and a min at $\left(\frac{3\pi}{4}, \frac{3\pi\sqrt{2}}{8} - \sqrt{2}\right)$.

3 Locate the local extrema of $y = (x^2 - 8)^{2/3}$ with the first derivative test. **Local mins at $(-2\sqrt{2},\, 0)$ and $(2\sqrt{2},\, 0)$; a local max at $(0,\, 4)$.**

Same basic steps as problems 1 and 2, but abbreviated a bit.

1. Find the derivative.

$$y = (x^2 - 8)^{2/3}$$

$$y' = \frac{2}{3}(x^2 - 8)^{-1/3}(2x) = \frac{4x}{3\sqrt[3]{x^2 - 8}}$$

2. Find the critical numbers.

   a. $\dfrac{4x}{3\sqrt[3]{x^2 - 8}} = 0$

   $$x = 0$$

   b. The first derivative will be undefined when the denominator is zero, so

   $$3\sqrt[3]{x^2 - 8} = 0$$
   $$\sqrt[3]{x^2 - 8} = 0$$
   $$x^2 - 8 = 0$$
   $$x^2 = 8$$
   $$x = \pm 2\sqrt{2}$$

   The critical numbers are $-2\sqrt{2}$, 0, and $2\sqrt{2}$.

3. Test the values.

$$y'(-10) = \frac{2}{3}\big((-10)^2 - 8\big)^{-1/3}\big(2 \cdot (-10)\big) \qquad y'(-1) = \frac{2}{3}\big((-1)^2 - 8\big)^{-1/3}\big(2 \cdot (-1)\big)$$

$$= \frac{2}{3}(positive)^{-1/3} \cdot negative \qquad\qquad = \frac{2}{3} \cdot (negative)^{-1/3} \cdot negative$$

$$= \frac{2}{3}\, positive \cdot negative \qquad\qquad\qquad = \frac{2}{3} \cdot negative \cdot negative$$

$$= negative \qquad\qquad\qquad\qquad\qquad = positive$$

$y'(1) = negative$ and $y'(10) = positive$ (What? I should do all your work?)

4. Make a sign graph (see the following figure).

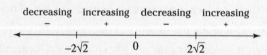

5. Find the y-values.

$$y = \left((-2\sqrt{2})^2 - 8\right)^{2/3} = 0 \text{ There's a local min at } (-2\sqrt{2},\, 0).$$

$$y = (0^2 - 8)^{2/3} = (-8)^{2/3} = 4 \text{ There's a local max at } (0, 4).$$

$$y = \left((2\sqrt{2})^2 - 8\right)^{2/3} = 0 \text{ There's another local min at } (2\sqrt{2},\, 0). \text{ Check out this interesting curve}$$
on your graphing calculator.

4   Using the first derivative test, determine the local extrema of $s = \dfrac{t^4 + 4}{-2t^2}$. **Local maxes at** $\left(-\sqrt{2}, -2\right)$ **and** $\left(\sqrt{2}, -2\right)$.

1. Do the differentiation thing.

$$s = \frac{t^4 + 4}{-2t^2}$$

$$s' = \frac{\left(t^4 + 4\right)'\left(-2t^2\right) - \left(t^4 + 4\right)\left(-2t^2\right)'}{\left(-2t^2\right)^2} = \frac{\left(4t^3\right)\left(-2t^2\right) - \left(t^4 + 4\right)\left(-4t\right)}{4t^4} = \frac{-t^4 + 4}{t^3}$$

2. Find the critical numbers.

$$\frac{-t^4 + 4}{t^3} = 0$$
$$4 - t^4 = 0$$
$$\left(2 - t^2\right)\left(2 + t^2\right) = 0$$
$$\left(\sqrt{2} - t\right)\left(\sqrt{2} + t\right)\left(2 + t^2\right) = 0$$
$$t = \sqrt{2} \ \text{ or } \ -\sqrt{2}$$

So $-\sqrt{2}$ and $\sqrt{2}$ are two critical numbers of s.

 $t = 0$ is a third important number because $t = 0$ makes the derivative's denominator equal zero, and so you need to include zero on your sign graph in order to define test regions. Note, however, that $t = 0$ is *not* a critical number of s because s is undefined at $t = 0$. And because there is no point on s at $t = 0$, there cannot be a local extremum at $t = 0$.

3. Test values: You're on your own.

4. Make a sign graph (see the following figure).

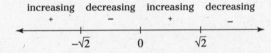

She loves me; she loves me not; she loves me; she loves me not.

5. Find the y-values.

$s\left(-\sqrt{2}\right) = \dfrac{\left(-\sqrt{2}\right)^4 + 4}{-2\left(-\sqrt{2}\right)^2} = \dfrac{4 + 4}{-4} = -2$ You climb up the hill to $\left(-\sqrt{2}, -2\right)$, then down, so there's a local max there.

$s(0) = \dfrac{0^4 + 4}{-2(0)^2} = undefined$ (which you already knew). Therefore, there's no local extremum at $t = 0$. Remember that if a problem asks you to identify only the x-values and not the y-values of the local extrema, and you only consider the sign graph, you would incorrectly conclude — using the current problem as an example — that there's a local min at $t = 0$. So you should always check where your function is undefined.

$s\left(\sqrt{2}\right) = \dfrac{\sqrt{2}^4 + 4}{-2\sqrt{2}^2} = -2$. Up then down again, so there's another local max at $\left(\sqrt{2}, -2\right)$.

As always, you should check out this function on your graphing calculator.

5 Use the second derivative test to analyze the critical numbers of the function from problem 2, $h(x) = \dfrac{x}{\sqrt{2}} + \cos x - \dfrac{\sqrt{2}}{2}$. **Local max at $x = -\dfrac{\sqrt{2}}{2}$; local min at $x = \dfrac{\sqrt{2}}{2}$.**

1. Find the second derivative.

$$h(x) = \frac{x}{\sqrt{2}} + \cos x - \frac{\sqrt{2}}{2}$$

$$h'(x) = \frac{1}{\sqrt{2}} - \sin x$$

$$h''(x) = -\cos x$$

2. Plug in the critical numbers (from problem 2).

$$h''\left(\frac{\pi}{4}\right) = -\cos\frac{\pi}{4} \qquad h''\left(\frac{3\pi}{4}\right) = -\cos\frac{3\pi}{4}$$

$$= -\frac{\sqrt{2}}{2} \qquad\qquad = \frac{\sqrt{2}}{2}$$

You're done. h is concave down at $x = -\dfrac{\sqrt{2}}{2}$, so there's a local max there, and h is concave up at $x = \dfrac{\sqrt{2}}{2}$, so there's a local min at that x-value.

(In problem 2, you already determined the y-values for these extrema.) h is an example of a function where the second derivative test is quick and easy.

6 Find the local extrema of $f(x) = -2x^3 + 6x^2 + 1$ with the second derivative test. **Local min at (0, 1); local max at (2, 9).**

1. Find the critical numbers.

$$f(x) = -2x^3 + 6x^2 + 1$$
$$f'(x) = -6x^2 + 12x$$
$$0 = -6x^2 + 12x$$
$$0 = -6x(x - 2)$$
$$x = 0,\ 2$$

2. Find the second derivative.

$$f'(x) = -6x^2 + 12x$$
$$f''(x) = -12x + 12$$

3. Plug in the critical numbers.

$$f''(0) = -12(0) + 12 \qquad\qquad f''(2) = -12(2) + 12$$
$$= 12 \ \text{(concave up: min)} \qquad\qquad = -12 \ \text{(concave down: max)}$$

4. Determine the y-coordinates for the extrema.

$$f(0) = -2(0)^3 + 6(0)^2 + 1 \qquad f(2) = -2(2)^3 + 6(2)^2 + 1$$
$$= 1 \qquad\qquad\qquad\qquad = 9$$

So there's a min at (0, 1) and a max at (2, 9). f is another function where the second derivative test works like a charm.

7 Find the local extrema of $y = 2x^4 - \frac{1}{3}x^6$ with the second derivative test. **You find local maxes at $x = -2$ and $x = 2$ with the second derivative test; you find a local min at $x = 0$ with street smarts.**

1. Find the critical numbers.

$$y = 2x^4 - \frac{1}{3}x^6$$
$$y' = 8x^3 - 2x^5$$
$$8x^3 - 2x^5 = 0$$
$$2x^3(4 - x^2) = 0$$
$$2x^3(2 - x)(2 + x) = 0$$
$$x = 0, \ 2, \ -2$$

2. Get the second derivative.

$$y' = 8x^3 - 2x^5$$
$$y'' = 24x^2 - 10x^4$$

3. Plug in.

$$y''(-2) = 24(-2)^2 - 10(-2)^4 \qquad y''(0) = 24(0)^2 - 10(0)^4 \qquad y''(2) = 24(2)^2 - 10(2)^4$$
$$= 96 - 160 \qquad\qquad\qquad = 0 \quad \text{thus inconclusive} \qquad = \text{same as } y''(-2)$$
$$= negative \ \text{thus a max} \qquad\qquad\qquad\qquad\qquad\qquad\qquad = negative \ \text{thus a max}$$

The second derivative test fails at $x = 0$, so you've got to use the first derivative test for that critical number. And this means, basically, that the second derivative test was a waste of time for this function.

TIP

If — as in the function for this problem — one of the critical numbers is $x = 0$, and you can see that the second derivative will equal zero at $x = 0$ (because, for example, all the terms of the second derivative will be simple powers of x), then the second derivative test will fail for $x = 0$ and thus, will likely be a waste of time. Use the first derivative test instead.

However, because this is a continuous function and because there's only one critical number between the two maxes you found, the only possibility is that there's a min at $x = 0$. Try this streetwise logic out on your teacher and let me know if it works.

8 Consider the function from problem 3, $y = (x^2 - 8)^{2/3}$, and the function $s = 8 + \frac{21t}{4} - \frac{7t^3}{4}$. Which is easier to analyze with the second derivative test and why? For the function you pick, use the second derivative test to find its local extrema. $s = 8 + \frac{21t}{4} - \frac{7t^3}{4}$; **local min at (–1, 4.5) and a local max at (–1, 11.5).**

TIP

The second derivative test fails where the second derivative is undefined (in addition to failing where the second derivative equals zero).

To pick, look at the first derivative of each function:

$$y = (x^2 - 8)^{2/3} \qquad\qquad s = 8 + \frac{21t}{4} - \frac{7t^3}{4}$$
$$y' = \frac{2}{3}(x^2 - 8)^{-1/3}(2x) \qquad s' = \frac{21}{4} - \frac{21}{4}t^2$$
$$= \frac{4x}{3(x^2 - 8)^{1/3}}$$

Do you see the trouble you're going to run into with $y(x)$? The first derivative is undefined at $x = \pm 2\sqrt{2}$. And the second derivative will also be undefined at those x-values, because when you take the second derivative with the quotient rule, squaring the bottom, the denominator will contain that same factor, $(x^2 - 8)$. The second derivative test will thus fail at $\pm 2\sqrt{2}$, and you'll have to use the first derivative test. In contrast to $y(x)$, the second derivative test works great with $s(t)$:

1. Get the critical numbers.

$$s' = \frac{21}{4} - \frac{21}{4}t^2$$
$$0 = \frac{21}{4} - \frac{21}{4}t^2$$
$$\frac{21}{4}t^2 = \frac{21}{4}$$
$$t = \pm 1$$

s' is not undefined anywhere, so -1 and 1 are the only critical numbers.

2. Do the second derivative.

$$s' = \frac{21}{4} - \frac{21}{4}t^2$$
$$s'' = -\frac{21}{2}t$$

3. Plug in the critical numbers.

$$s''(-1) = \frac{21}{2} \quad \text{(concave up: min)}$$
$$s''(1) = -\frac{21}{2} \quad \text{(concave down: max)}$$

4. Get the heights of the extrema.

$$s(-1) = 8 + \frac{21(-1)}{4} - \frac{7(-1)^3}{4} = 4.5$$
$$s(1) = 8 + \frac{21(1)}{4} - \frac{7(1)^3}{4} = 11.5$$

You're done. s has a local min at $(-1, 4.5)$ and a local max at $(-1, 11.5)$.

9 Find the absolute extrema of $f(x) = \sin x + \cos x$ on the interval $[0, 2\pi]$. **Absolute max at** $\left(\frac{\pi}{4}, \sqrt{2}\right)$**; absolute min at** $\left(\frac{5\pi}{4}, -\sqrt{2}\right)$.

1. Find critical numbers.

$$f(x) = \sin x + \cos x$$
$$f'(x) = \cos x - \sin x$$
$$0 = \cos x - \sin x$$
$$\sin x = \cos x \quad \text{(divide both sides by } \cos x)$$
$$\tan x = 1$$
$$x = \frac{\pi}{4}, \frac{5\pi}{4} \quad \text{(the solutions in the given interval)}$$

The derivative is never undefined, so these are the only critical numbers.

If you divide both sides of an equation by something that can equal zero at one or more x-values (like you do above when dividing both sides by $\cos x$), you may miss one or more solutions. You have to check whether any of those x-values is a solution. In this problem, $\cos x = 0$ at $\frac{\pi}{2}$ and $\frac{3\pi}{2}$, and it's easy to check (in line 4 above) that $\sin x$ does not equal $\cos x$ at either of those values. If it did, you'd have one or two more solutions and one or two more critical numbers. Note that you have to check these values before dividing out the "dividing thing" ($\cos x$ here).

2. Evaluate the function at the critical numbers.

$$f\left(\frac{\pi}{4}\right) = \sin\frac{\pi}{4} + \cos\frac{\pi}{4} \qquad f\left(\frac{5\pi}{4}\right) = \sin\frac{5\pi}{4} + \cos\frac{5\pi}{4}$$

$$= \frac{\sqrt{2}}{2} + \frac{\sqrt{2}}{2} \qquad\qquad = -\frac{\sqrt{2}}{2} - \frac{\sqrt{2}}{2}$$

$$= \sqrt{2} \qquad\qquad\qquad = -\sqrt{2}$$

3. Evaluate the function at the endpoints of the interval.

$$f(0) = \sin 0 + \cos 0 = 1$$
$$f(2\pi) = \sin 2\pi + \cos 2\pi = 1$$

4. The largest of the four answers from Steps 2 and 3 is the absolute max; the smallest is the absolute min.

The absolute max is at $\left(\frac{\pi}{4}, \sqrt{2}\right)$. The absolute min is at $\left(\frac{5\pi}{4}, -\sqrt{2}\right)$.

10 Find the absolute extrema of $g(x) = 2x^3 - 3x^2 - 5$ on the interval $[-.5, .5]$. **Absolute min at (−.5, −6); absolute max at (0, −5).**

1. Find critical numbers.

$$g(x) = 2x^3 - 3x^2 - 5$$
$$g'(x) = 6x^2 - 6x$$
$$0 = 6x^2 - 6x$$
$$0 = 6x(x - 1)$$
$$x = 0, 1$$

$x = 1$ is neglected because it's outside the given interval; $x = 0$ is your only critical number.

2. Evaluate the function at $x = 0$.

$$g(0) = 2(0)^3 - 3(0)^2 - 5 = -5$$

3. Do the endpoint thing.

$$g(-.5) = 2(-.5)^3 - 3(-.5)^2 - 5$$
$$= 2 \cdot (-.125) - 3 \cdot .25 - 5$$
$$= -6$$

$$g(.5) = 2 \cdot .5^3 - 3 \cdot .5^2 - 5$$
$$= 2 \cdot .125 - 3 \cdot .25 - 5$$
$$= -5.5$$

4. Pick the smallest and largest answers from Steps 2 and 3.

The absolute min is at the left endpoint, (−.5, −6). The absolute max is smack dab in the middle, (0, −5).

11 Find the absolute extrema of $p(x) = (x+1)^{4/5} - .5x$ on the interval $[-2, 31]$. **Absolute max at (–2, 2); absolute mins at (–1, .5) and (31, .5).**

I think you know the steps by now.

$$p(x) = (x+1)^{4/5} - .5x$$

$$p'(x) = \frac{4}{5}(x+1)^{-1/5} - .5$$

$$= \frac{4}{5(x+1)^{1/5}} - .5$$

$$0 = \frac{4}{5(x+1)^{1/5}} - .5$$

$$.5 = \frac{4}{5(x+1)^{1/5}}$$

$$2.5(x+1)^{1/5} = 4$$

$$(x+1)^{1/5} = \frac{8}{5}$$

$$(x+1) = \left(\frac{8}{5}\right)^5$$

$$x = 9.48576$$

That's one critical number, but $x = -1$ is also one because it produces an undefined derivative.

$$p(-1) = (-1+1)^{4/5} - .5(-1)$$

$$= .5$$

$$p(9.48576) = (9.48576 + 1)^{4/5} - .5(9.48576)$$

$$= 1.81072$$

Left endpoint: $p(-2) = (-2+1)^{4/5} - .5(-2) = 2$

Right endpoint: $p(31) = (31+1)^{4/5} - .5(31) = 16 - 15.5 = .5$

Your absolute max is at the left endpoint: (–2, 2). There's a tie for the absolute min: At the cusp: (–1, .5) and at the right endpoint: (31, .5).

12 Find the absolute extrema of $q(x) = 2\cos 2x + 4\sin x$ on the interval $\left[-\frac{\pi}{2}, \frac{5\pi}{4}\right]$. **Absolute min at $\left(-\frac{p}{2}, -6\right)$; absolute maxes at $\left(\frac{p}{6}, 3\right)$ and $\left(\frac{5p}{6}, 3\right)$.**

$$q(x) = 2\cos 2x + 4\sin x$$

$$q'(x) = -2\sin 2x \cdot 2 + 4\cos x$$

$$0 = -4\sin 2x + 4\cos x$$

$$0 = \sin 2x - \cos x \quad (\text{dividing by} -4)$$

$$0 = 2\sin x \cos x - \cos x \quad (\text{trig identity})$$

$$0 = \cos x (2\sin x - 1)$$

$$\cos x = 0 \qquad 2\sin x - 1 = 0$$

$$x = \frac{-\pi}{2}, \frac{\pi}{2} \quad \text{or} \quad \sin x = \frac{1}{2}$$

$$x = \frac{\pi}{6}, \frac{5\pi}{6}$$

REMEMBER

Technically $x = \frac{-\pi}{2}$ is not one of the critical numbers; being at an endpoint, it is refused membership in the critical number club. It's a moot point though, because you have to evaluate the endpoints anyway.

$$q\left(\frac{\pi}{6}\right) = 2\cos\left(2 \cdot \frac{\pi}{6}\right) + 4\sin\frac{\pi}{6}$$
$$= 2 \cdot \frac{1}{2} + 4 \cdot \frac{1}{2} = 3$$
$$q\left(\frac{\pi}{2}\right) = 2\cos\left(2 \cdot \frac{\pi}{2}\right) + 4\sin\frac{\pi}{2}$$
$$= -2 + 4 = 2$$
$$q\left(\frac{5\pi}{6}\right) = 2\cos\left(2 \cdot \frac{5\pi}{6}\right) + 4\sin\frac{5\pi}{6}$$
$$= 2 \cdot \frac{1}{2} + 4 \cdot \frac{1}{2} = 3$$

Left endpoint: $q\left(-\frac{\pi}{2}\right) = 2\cos\left(2 \cdot -\frac{\pi}{2}\right) + 4\sin\left(-\frac{\pi}{2}\right) = -2 + 4(-1) = -6$

Right endpoint: $q\left(\frac{5\pi}{4}\right) = 2\cos\left(2 \cdot \frac{5\pi}{4}\right) + 4\sin\frac{5\pi}{4} = 2 \cdot 0 + 4\left(-\frac{\sqrt{2}}{2}\right) \approx -2.828$

Pick your winners: Absolute min at left endpoint: $\left(-\frac{\pi}{2}, -6\right)$ and a tie for absolute max: $\left(\frac{\pi}{6}, 3\right)$ and $\left(\frac{5\pi}{6}, 3\right)$.

13 Find the intervals of concavity and the inflection points of $f(x) = -2x^3 + 6x^2 - 10x + 5$. *f* **is concave up from** $-\infty$ **to the inflection point at (1, –1), then concave down from there to** ∞.

1. Get the second derivative.

$$f(x) = -2x^3 + 6x^2 - 10x + 5$$
$$f'(x) = -6x^2 + 12x - 10$$
$$f''(x) = -12x + 12$$

2. Set equal to 0 and solve.

$$-12x + 12 = 0$$
$$x = 1$$

3. Check for *x*-values where the second derivative is undefined. None.

4. Test your two regions — to the left and to the right of $x = 1$ — and make your sign graph (see the following figure).

$$f''(x) = -12x + 12$$

$$f''(0) = 12$$
$$f''(2) = -12$$

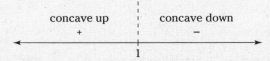

concave up   concave down
+   −
1

Because the concavity switches at $x = 1$ and because f'' equals zero there, there's an inflection point at $x = 1$.

5. Find the height of the inflection point.

$$f(x) = -2x^3 + 6x^2 - 10x + 5$$
$$f(1) = -1$$

Thus f is concave up from $-\infty$ to the inflection point at $(1, -1)$, then concave down from there to ∞. As always, you should check your result on your graphing calculator. *Hint:* To get a good feel for the look of this function, you'll need a fairly odd graphing window — try something like *xmin* = -2, *xmax* = 4, *ymin* = -20, *ymax* = 20.

14 Find the intervals of concavity and the inflection points of $g(x) = x^4 - 12x^2$. **g is concave up from $-\infty$ to the inflection point at $\left(-\sqrt{2},\ -20\right)$; then concave down to an inflection point at $\left(\sqrt{2},\ -20\right)$; then concave up again to ∞.**

1. Find the second derivative.

$$g(x) = x^4 - 12x^2$$
$$g'(x) = 4x^3 - 24x$$
$$g''(x) = 12x^2 - 24$$

2. Set to 0 and solve.

$$12x^2 - 24 = 0$$
$$x^2 = 2$$
$$x = \pm\sqrt{2}$$

3. Is the second derivative undefined anywhere? No.

4. Test the three regions and make a sign graph. See the following figure.

$$g''(x) = 12x^2 - 24$$

$$g''(-2) = 24$$
$$g''(0) = -24$$
$$g''(2) = 24$$

Because the concavity switched signs at the two zeros of g'', there are inflection points at these two x-values.

5. Find the heights of the inflection points.

$$g(x) = x^4 - 12x^2$$

$$g\left(-\sqrt{2}\right) = -20$$
$$g\left(\sqrt{2}\right) = -20$$

g is concave up from $-\infty$ to the inflection point at $\left(-\sqrt{2},\ -20\right)$, concave down from there to another inflection point at $\left(\sqrt{2},\ -20\right)$, then concave up again from there to ∞.

15 Find the intervals of concavity and the inflection points of $p(x) = \frac{x}{x^2 + 9}$. **Concave down from**
∞ **to an inflection point at** $\left(-3\sqrt{3}, -\frac{\sqrt{3}}{12}\right)$, **then concave up till the inflection point at** $(0, 0)$,
then concave down again till the third inflection point at $\left(3\sqrt{3}, \frac{\sqrt{3}}{12}\right)$, **and, finally, concave**
up to ∞.

1. Get the second derivative.

$$p'(x) = \frac{(x)'(x^2 + 9) - (x)(x^2 + 9)'}{(x^2 + 9)^2} \qquad p''(x) = \frac{(9 - x^2)'(x^2 + 9)^2 - (9 - x^2)\left((x^2 + 9)^2\right)'}{(x^2 + 9)^4}$$

$$= \frac{x^2 + 9 - 2x^2}{(x^2 + 9)^2} \qquad\qquad = \frac{-2x(x^2 + 9)^2 - (9 - x^2)2(x^2 + 9)2x}{(x^2 + 9)^4}$$

$$= \frac{9 - x^2}{(x^2 + 9)^2} \qquad\qquad = \frac{(x^2 + 9)\left[-2x(x^2 + 9) - 4x(9 - x^2)\right]}{(x^2 + 9)^4}$$

$$\qquad\qquad = \frac{-2x^3 - 18x - 36x + 4x^3}{(x^2 + 9)^3}$$

$$\qquad\qquad = \frac{2x(x^2 - 27)}{(x^2 + 9)^3}$$

2. Set equal to 0 and solve.

$$\frac{2x(x^2 - 27)}{(x^2 + 9)^3} = 0$$

$$2x(x^2 - 27) = 0$$

$$2x = 0 \qquad x^2 - 27 = 0$$
$$x = 0 \quad \text{or} \quad x \pm 3\sqrt{3}$$

3. Check for undefined points of the second derivative. None.

4. Test four regions with the second derivative. You can skip the sign graph. *Tip:* You can do all
   of this in your head because all that matters is whether the answers are positive or negative.

$$p''(x) = \frac{2x(x^2 - 27)}{(x^2 + 9)^3}$$

$$p''(-10) = \frac{2(-10)\left((-10)^2 - 27\right)}{\left((-10)^2 + 9\right)^3} \qquad\qquad p''(-1) = \frac{2(-1)\left((-1)^2 - 27\right)}{\left((-1)^2 + 9\right)^3}$$

$$= \frac{2(N)(P)}{P^3} \qquad\qquad\qquad = \frac{2(N)(N)}{P^3}$$

$$= \frac{N}{P} \qquad\qquad\qquad = \frac{P}{P}$$

$$= N \qquad\qquad\qquad = P$$

$$p''(1) = \frac{2(1)\left((1)^2 - 27\right)}{\left((1)^2 + 9\right)^3} \qquad\qquad p''(10) = \frac{2(10)\left((10)^2 - 27\right)}{\left((10)^2 + 9\right)^3}$$

$$= \frac{2(P)(N)}{P^3} \qquad\qquad\qquad = \frac{2(P)(P)}{P^3}$$

$$= \frac{N}{P} \qquad\qquad\qquad = \frac{P}{P}$$

$$= N \qquad\qquad\qquad = P$$

The concavity goes *N, P, N, P* so there's an inflection point at each of the three zeros of p''.

5. Find the heights of the inflection points.

$$p(x) = \frac{x}{x^2+9}$$

$$p\left(-3\sqrt{3}\right) = \frac{-3\sqrt{3}}{\left(-3\sqrt{3}\right)^2+9} \qquad p(0)=0 \qquad p\left(3\sqrt{3}\right) = \frac{3\sqrt{3}}{\left(3\sqrt{3}\right)^2+9}$$

$$= \frac{-3\sqrt{3}}{27+9} \qquad\qquad\qquad = \frac{\sqrt{3}}{12}$$

$$= \frac{-\sqrt{3}}{12}$$

Taking a drive on highway p, you'll be turning right from negative ∞ to $\left(-3\sqrt{3}, -\frac{\sqrt{3}}{12}\right)$, where you'll point straight ahead for an infinitesimal moment, then you'll be turning left till $(0, 0)$, then right again till $\left(3\sqrt{3}, \frac{\sqrt{3}}{12}\right)$, and on your final leg to ∞, you'll round a *very long* bend to the left.

16 Find the intervals of concavity and the inflection points of $q(x) = \sqrt[5]{x} - \sqrt[3]{x}$. **Concave down from $-\infty$ till an inflection point at about $(-.085, -.171)$, then concave up till a vertical inflection point at $(0, 0)$, then concave down till a third inflection point at about $(.085, .171)$, then concave up out to ∞.**

You know the routine.

$$q(x) = \sqrt[5]{x} - \sqrt[3]{x}$$

$$q'(x) = \frac{1}{5} x^{-4/5} - \frac{1}{3} x^{-2/3}$$

$$q''(x) = -\frac{4}{25} x^{-9/5} + \frac{2}{9} x^{-5/3}$$

$$\frac{-4}{25x^{9/5}} + \frac{2}{9x^{5/3}} = 0$$

Whoops, I guess this algebra's kind of messy. Let's get the zeros on our calculators: just graph $y = \frac{-4}{25x^{9/5}} + \frac{2}{9x^{5/3}}$ and find the x-intercepts. There are two: $x \approx -0.085$ and $x \approx 0.085$.

So you've got two "critical numbers," right? Wrong! Don't forget to check for undefined points of the second derivative. Because $q''(x) = -\frac{4}{25} x^{-9/5} + \frac{2}{9} x^{-5/3}$, q'' is undefined at $x = 0$. Since $q(x)$ *is* defined at $x = 0$, 0 is another critical number. So you have three critical numbers and four regions. You can test them with -1, $-.01$, $.01$, and 1:

$$q''(x) = -\frac{4}{25} x^{-9/5} + \frac{2}{9} x^{-5/3}$$

$$q''(-1) = -\frac{14}{225} \qquad q''(-.01) \approx 158 \qquad q''(.01) \approx -158 \qquad q''(1) = \frac{14}{225}$$

Thus the concavity goes: down, up, down, up. Because the second derivative is zero at -0.085 and 0.085 and because the concavity switches there, you can conclude that there are inflection points at those two x-values. But because both the first and second derivatives are undefined at $x = 0$, you have to check whether there's a vertical tangent there. You can see that there is by just looking at the graph, but if want to be rigorous about it, you figure the limit of the first derivative as x approaches zero. Since that equals infinity, you've got a vertical tangent at $x = 0$, and thus there's an inflection point there.

Now plug in -0.085, 0, and 0.085 into q to get the y-values and you're done.

17 For $g(x) = x^3 + x^2 - x$, find all the values c in the interval $(-2, 1)$ that satisfy the Mean Value Theorem. **The values of c are** $\dfrac{-1 - \sqrt{7}}{3}$ **and** $\dfrac{-1 + \sqrt{7}}{3}$.

1. Find the first derivative.

$$g(x) = x^3 + x^2 - x$$
$$g'(x) = 3x^2 + 2x - 1$$

2. Figure the slope between the endpoints of the interval.

$$g(-2) = (-2)^3 + (-2)^2 - (-2) \qquad m = \frac{g(-2) - g(1)}{-2 - 1}$$
$$= -2$$
$$= \frac{-2 - 1}{-2 - 1}$$
$$g(1) = 1$$
$$= 1$$

3. Set the derivative equal to this slope and solve.

$$3x^2 + 2x - 1 = 1 \qquad x = \frac{-2 \pm \sqrt{4 - (-24)}}{6}$$
$$3x^2 + 2x - 2 = 0$$
$$= \frac{-2 \pm 2\sqrt{7}}{6}$$
$$= \frac{-1 - \sqrt{7}}{3} \text{ or } \frac{-1 + \sqrt{7}}{3}$$

Both are inside the given interval, so you've got two answers.

18 For $s(t) = t^{4/3} - 3t^{1/3}$, find all the values c in the interval $(0, 3)$ that satisfy the Mean Value Theorem. **The value of c is ¾.**

1. Find the first derivative.

$$s(t) = t^{4/3} - 3t^{1/3}$$
$$s'(t) = \frac{4}{3}t^{1/3} - t^{-2/3}$$

2. Figure the slope between the endpoints of the interval.

$$s(0) = 0 \qquad\qquad m = \frac{s(3) - s(0)}{3 - 0}$$
$$s(3) = 3^{4/3} - 3 \cdot 3^{1/3}$$
$$= \frac{0 - 0}{3}$$
$$= 0$$
$$= 0$$

3. Set the derivative equal to the result from Step 2 and solve.

$$\frac{4}{3}t^{1/3} - t^{-2/3} = 0$$
$$t^{-2/3}\left(\frac{4}{3}t^1 - 1\right) = 0$$
$$t^{-2/3} = 0 \text{ or } \frac{4}{3}t^1 - 1 = 0$$
$$\varnothing \qquad \text{ or } t = \frac{3}{4}$$

Graph s and check that its slope at $t = \dfrac{3}{4}$ is zero.

Chapter 8

Using Differentiation to Solve Practical Problems

··

In This Chapter

▶ Optimizing space

▶ Relating rates

▶ Getting up to speed with position, velocity, and acceleration

▶ Going off on a tangent

▶ Doing $\sqrt{37}$ in your head

··

*N*ow that you're an expert at finding derivatives, I'm sure you can't wait to put your expertise to use solving some practical problems. In this section, you find problems that actually come up in the real world — problems like how should a cat rancher use 200 feet of fencing to build a three-sided corral next to a river (he only needs three sides because the river makes the fourth side and cats hate water) to maximize the grazing area for his cats?

Optimization Problems: From Soup to Nuts

Optimization problems are one of the most practical types of calculus problems. You use the techniques shown below whenever you want to maximize or minimize something: like maximizing profit or area or volume or minimizing cost or energy consumption, and so on.

Q. A rancher has 400 feet of fencing and wants to build a corral that's divided into three equal rectangles. See the following figure. What length and width will maximize the area?

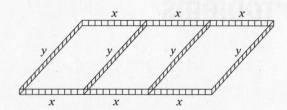

A. 100 feet by 50 feet with an area of 5000 square feet.

1. **Draw a diagram and label with variables.**

2. a. **Express the thing you want maximized, the area, as a function of the variables.**

   $Area = Length \times Width$

   $A = 3x \cdot y$

   b. **Use the given information to relate the two variables to each other.**

   $6x + 4y = 400$

   $3x + 2y = 200$

   c. **Solve for one variable and substitute into the equation from Step 2a to create a function of a single variable.**

   $2y = 200 - 3x$

   $y = 100 - 1.5x$

   $A = 3x \cdot y$

   $A(x) = 3x(100 - 1.5x)$

   $= 300x - 4.5x^2$

3. **Determine the domain of the function.**

   You can't have a negative length of fence, so x can't be negative. And if you build the ridiculous corral with no width, all 400 feet of fencing would equal $6x$. So

   $x \geq 0$ and $6x \leq 400$

   $x \leq \dfrac{200}{3}$

4. **Find the critical numbers of $A(x)$.**

   $A(x) = 300x - 4.5x^2$

   $A'(x) = 300 - 9x$

   $0 = 300 - 9x$

   $9x = 300$

   $x = \dfrac{100}{3}$

   $A'(x)$ is defined everywhere, so $\dfrac{100}{3}$ is the only critical number.

5. **Evaluate $A(x)$ at the critical number and at the endpoints of the domain.**

   $A(0) = 0$

   $A\left(\dfrac{100}{3}\right) = 300\left(\dfrac{100}{3}\right) - 4.5\left(\dfrac{100}{3}\right)^2$

   $= 5000$

   $A\left(\dfrac{200}{3}\right) = 0$

   The first and third results above should be obvious because they represent corrals with zero length and zero width. You're done. $x = \dfrac{100}{3}$ will maximize the area. Plug that into $y = 100 - 1.5x$ and you get $y = 50$. So the largest corral is $3 \cdot \dfrac{100}{3}$, or 100 feet long, 50 feet wide, and has an area of 5000 square feet.

1. What are the dimensions of the soup can of greatest volume that can be made with 50 square inches of tin? (The entire can, including the top and bottom, are made of tin.) And what's its volume?

Solve It

2. A Norman window is in the shape of a semicircle above a rectangle. If the straight edges of the frame cost $20 per linear foot and the circular frame costs $25 per linear foot and you want a window with an area of 20 square feet, what dimensions will minimize the cost of the frame?

Solve It

3. A right triangle is placed in the first quadrant with its legs on the x and y axes. Given that its hypotenuse must pass through the point (2, 5), what are the dimensions and area of the smallest such triangle?

Solve It

4. You're designing an open-top cardboard box for a purveyor of nuts. The top will be made of clear plastic, but the plastic-box-top designer is handling that. The box must have a square base and two cardboard pieces that divide the box into four sections for the almonds, cashews, pecans, and walnuts. See the following figure. Given that you want a box with a volume of 72 cubic inches, what dimensions will minimize the total cardboard area and thus minimize the cost of the cardboard? What's the total area of cardboard?

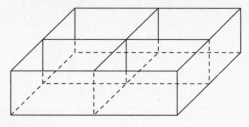

Solve It

Problematic Relationships: Related Rates

Related rates problems are the Waterloo for many a calculus student. But they're not that bad after you get the basic technique down. The best way to learn them is by working through examples, so get started!

After working each problem, ask yourself whether the answer makes sense. Asking this question is one of the best things you can do to increase your success in mathematics and science. And while it's not always possible to decide whether a math answer is reasonable, when it's possible, this inquiry should be a quick, extra step of every problem you do.

Q. A homeowner decides to paint his home. He picks up a home improvement book, which recommends that a ladder should be placed against a wall such that the distance from the foot of the ladder to the bottom of the wall is one third the length of the ladder. Not being the sharpest tool in the shed, the homeowner gets mixed up and thinks that it's the distance from the *top* of the ladder to the base of the wall that should be a third of the ladder's length. He sets up his 18 foot ladder accordingly, and — despite this unstable ladder placement — he manages to climb the ladder and start painting. (Perhaps the foot of the ladder is caught on a tree root or something.) His luck doesn't last long, and the ladder begins to slide rapidly down the wall. One foot before the top of the ladder hits the ground, it's falling at a rate of *20 feet/second*. At this moment, how fast is the foot of the ladder moving away from the wall?

A. Roughly 1.11 *feet/second*.

1. Draw a diagram, labeling it with any *unchanging* measurements and assigning variables to any *changing* things. See the following figure.

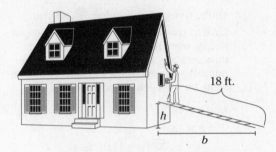

You don't have to draw the house — the basic triangle is enough. But I've sketched a fuller picture of this scenario to make clear what a bonehead this guy is.

2. List all given rates and the rate you're asked to figure out. Write these rates as derivates with respect to time.

You're told that the ladder is *falling* at a rate of 20 *ft / sec*. Going down is *negative*, so

$$\frac{dh}{dt} = -20 \qquad \frac{db}{dt} = ?$$

(*h* is for the distance from the top of the ladder to the bottom of the wall; *b* is for the distance from the base of the ladder to the wall.)

3. Write down the formula that connects the variables in the problem, *h* and *b*.

That's the Pythagorean Theorem, of course:

$a^2 + b^2 = c^2$, thus

$h^2 + b^2 = 18^2$

continued

4. Differentiate with respect to time.

This is a lot like implicit differentiation because you're differentiating with respect to t but the equation is in terms of h and b.

$$h^2 + b^2 = 18^2$$

$$2h\frac{dh}{dt} + 2b\frac{db}{dt} = 0$$

5. Substitute known values for the rates and variables in the equation from Step 4, and then solve for the thing you're asked to determine.

You're trying to determine $\frac{db}{dt}$, so you have to plug numbers into everything else. But, as often happens, you don't have a number for b, so use a formula to get the number you need. This will usually be the same formula you already used.

$$h^2 + b^2 = 18^2$$
$$1^2 + b^2 = 18^2$$

$$b = \pm\sqrt{323} \approx \pm17.97 \text{ feet}$$

(Obviously, you can reject the negative answer.)

Now you've got what you need to finish the problem.

$$2h\frac{dh}{dt} + 2b\frac{db}{dt} = 0$$

$$2(1)(-20) + 2(17.97)\frac{db}{dt} = 0$$

$$\frac{db}{dt} = \frac{40}{35.94} \approx 1.11 \text{ feet/sec}$$

6. Ask yourself whether your answer is reasonable.

Yes, it does make sense. Lean a yardstick against a wall so the bottom of it is about 4 or 5 inches from the wall. Then push the bottom of the yardstick the 4 or 5 inches to the wall. You'll see that the top would barely move up. Right triangles with a fixed hypotenuse like this one always work like that. If one leg is much shorter than the other, the short leg can change a lot while the long leg barely changes. It's a by-product of the Pythagorean Theorem.

5. A farmer's hog trough is 10 feet long, and its cross-section is an isosceles triangle with a base of 2 feet and a height of 2 feet 6 inches (with the vertex at the bottom, naturally). The farmer is pouring swill into the trough at a rate of 1 *cubic foot per minute*. Just as the swill reaches the brim, her three hogs start violently sucking down the swill at a rate of ½ *cubic foot per minute* for each hog. They're going at it so vigorously that another ½ cubic foot of swill is being splashed out of the trough each minute. The farmer keeps pouring in swill, but she's no match for his hogs. When the depth of the swill falls to 1 foot 8 inches, how fast is the swill level falling?

Solve It

6. A pitcher delivers a fastball, which the batter pops up — it goes straight up above home plate. When it reaches a height of 60 feet, it's moving up at a rate of 50 *ft/sec*. At this point, how fast is the distance from the ball to 2nd base growing? *Note:* The distance between the bases of a baseball diamond is 90 feet.

Solve It

7. A six-foot tall man looking over his shoulder sees his shadow that's cast by a 15-foot tall lamp post in front of him. The shadow frightens him so he starts running away from it — toward the lamp post. Unfortunately, this only makes matters worse, as it causes the frightening head of the shadow to gain on him. He starts to panic and runs even faster. Five feet before he crashes into the lamp post, he's running at a speed of 15 *miles/hour.* At this point, how fast is the tip of the shadow moving?

Solve It

8. Salt is being unloaded onto a conical pile at a rate of 200 *cubic feet per minute.* If the height of the cone-shaped pile is always equal to the radius of the cone's base, how fast is the height of the pile increasing when it's 18 feet tall?

Solve It

A Day at the Races: Position, Velocity, and Acceleration

The most important thing to know about this type of problem is that velocity is the derivative of position and acceleration is the derivative of velocity. The following points about position, velocity, and acceleration with regard to the chariot race in Figure 8-1 provide some keys to approaching these problems.

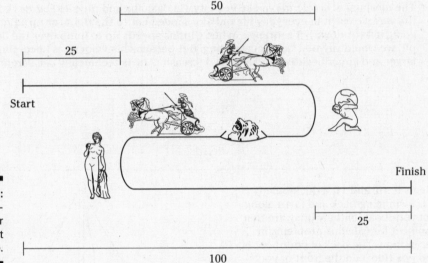

Figure 8-1:
A 200-palameter chariot race.

- The finish is 100 palameters from the start *as the crow flies*, so 100 palameters is the total *displacement*. (A *palameter* is a little-known unit of distance used in ancient Rome equal to the length of Julius Caesar's palace — roughly 380 feet.) Say the start is at (0, 0) on a coordinate system and the finish is at (100, 0). It's 100 from 0 to 100, of course, so 100 is the total displacement.

- *Distance* is different. You can see that the charioteers backtrack 50 palameters in the middle of the race. Because there are two extra 50-palameter legs, the total length of the race is 200 palameters — that's the distance. Distance is always positive or zero.

- Displacement to the *left* is *negative* (in other problems, down would be negative). When, say, Maximus passes Atlas, his *position* is 75 palameters from the start. At Aphrodite, he's back to only 25 palameters from the start as the crow flies. Displacement equals final position minus initial position, so from Atlas to Aphrodite is a displacement of 25 – 75, or –50 palameters.

- *Velocity* is related to displacement, not distance traveled. Velocity has a special meaning in calculus and physics so forget the everyday meaning of it. Like displacement, if you're going *left* (or *down*), that's a *negative* velocity. And here's a critical point: When you switch directions, your velocity is *zero*. Think of a ball thrown straight up. At its peak, for an infinitesimal moment, it is motionless, so its velocity is zero.

✔ *Average velocity* is defined as *total displacement* divided by *total time*. Say Glutius completes the race in 1 hour. Because he traveled 200 palameters, his *average speed* would be 200 *palameters per hour*. But because the total displacement is only 100, his average velocity would be a mere 100 *palameters per hour* (roughly 7 *miles per hour*).

✔ *Speed* is regular old speed, and, unlike velocity, it's always positive (or zero). If Maximus picks up speed to make the jump over the lion pit, his speed, naturally, would be increasing. *Note:* His velocity would be *decreasing* — even though we would see him speeding up — because his velocity would be negative and would be becoming a larger and larger negative.

✔ The meaning of *acceleration*, like velocity, (in calculus and physics) agrees with the way we use it in everyday life only for movement to the right (or up). But, going left (or down), it's strange. When Glutius speeds up to jump over the lion pit, we would say that he's accelerating. But because his velocity is becoming a larger and larger negative and is thus decreasing, he is technically *decelerating*.

9. For problems 9, 10, and 11, a duck-billed platypus is swimming back and forth along the side of your boat, blithely unaware that he's the subject for calculus problems in rectilinear motion. The back of your boat is at the zero position, and the front of your boat is in the positive direction (see the following figure). $s(t)$ gives his position (in feet) as a function of time (seconds). Find his a) position, b) velocity, c) speed, and d) acceleration, at t = 2 seconds.

Solve It

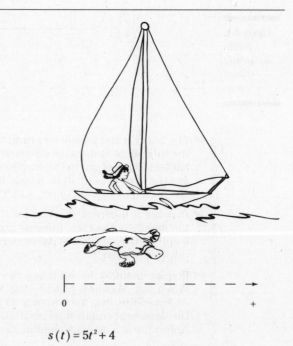

$$s(t) = 5t^2 + 4$$

10. $s(t) = 3t^4 - 5t^3 + t - 6$

Solve It

11. $s(t) = \frac{1}{t} + \frac{8}{t^3} - 3$

Solve It

12. For problems 12, 13, and 14, a three-toed
sloth is hanging onto a tree branch and
moving right and left along the branch.
(The tree trunk is at zero and the positive
direction goes out from the trunk.) $s(t)$
gives his position (in feet) as a function of
time (seconds). Between $t = 0$ and $t = 5$, for
each problem, find a) the intervals when
he's moving away from the trunk, the intervals when he's moving toward the trunk,
and when and where he turns around; b)
his total distance moved and his average
speed; and c) his total displacement and
his average velocity.

$s(t) = 2t^3 - t^2 + 8t - 5$

Solve It

13. $s(t) = t^4 + t^2 - t$

Solve It

14. $s(t) = \dfrac{t+1}{t^2+4}$

Solve It

Make Sure You Know Your Lines: Tangents and Normals

In everyday life, it's perfectly normal to go off on a tangent now and then. In calculus, on the other hand, there is nothing at all normal about a tangent. You need only note a couple points before you're ready to try some problems:

✔ At its point of tangency, a *tangent* line has the same slope as the curve it's tangent to. In calculus, whenever a problem involves slope, you should immediately think derivative. The derivative is the key to all tangent line problems.

✔ At its point of intersection to a curve, a *normal* line is *perpendicular* to the tangent line drawn at that same point. When any problem involves perpendicular lines, you use the rule that perpendicular lines have slopes that are opposite reciprocals. So all you do is use the derivative to get the slope of the tangent line, and then the opposite reciprocal of that gives you the slope of the normal line.

Ready to try a few problems? Say, that reminds me. I once had this problem with my carburetor. I took my car into the shop, and the mechanic told me the problem would be easy to fix, but when I went back to pick up my car . . . Wait a minute. Where was I?

Q. Find all lines through $(1, -4)$ either tangent to or normal to $y = x^3$. For each tangent line, give the point of tangency and the equation of the line; for the normal lines, give only the points of normalcy.

A. Point of tangency is $(2, 8)$; equation of tangent line is $y = 12x - 16$. Points of normalcy are approximately $(-1.539, -3.645)$, $(-.335, -.038)$, and $(.250, .016)$.

1. Find the derivative.

   $y = x^3$

   $y' = 3x^2$

2. For the tangent lines, set the slope from the general point $\left(x, x^3\right)$ to $(1, -4)$ equal to the derivative and solve.

   $$\frac{-4 - x^3}{1 - x} = 3x^2$$
   $$-4 - x^3 = 3x^2 - 3x^3$$
   $$2x^3 - 3x^2 - 4 = 0$$
   $$x = 2 \quad (\text{I used my calculator.})$$

3. Plug this solution into the original function to find the point of tangency.

   The point is $(2, 8)$.

4. Get your algebra fix by finding the equation of the tangent line that passes through $(1, -4)$ and $(2, 8)$.

   You can use either the point-slope form or the two-point form to arrive at $y = 12x - 16$.

5. For the normal lines, set the slope from the general point $\left(x, x^3\right)$ to $(1, -4)$ equal to the opposite reciprocal of the derivative and solve.

   $$\frac{-4 - x^3}{1 - x} = \frac{-1}{3x^2}$$
   $$-12x^2 - 3x^5 = x - 1$$
   $$3x^5 + 12x^2 + x - 1 = 0$$
   $$x \approx -1.539, -.335,$$
   $$\text{or } .250 \quad (\text{Use your}$$
   $$\text{calculator.})$$

6. Plug these solutions into the original function to find the points of normalcy.

   Plugging the points into $y = x^3$ gives you the three points: $(-1.539, -3.645)$, $(-.335, -.038)$, and $(.250, .016)$.

15. Two lines through the point (1, –3) are tangent to the parabola $y = x^2$. Determine the points of tangency.

Solve It

16. The Earth has a radius of 4,000 miles. Say you're standing on the shore and your eyes are 5' 3.36" above the surface of the water. How far out can you see to the horizon before the Earth's curvature makes the water dip below the horizon? See the following figure.

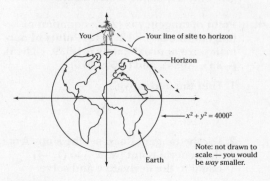

You — Your line of site to horizon

— Horizon

$x^2 + y^2 = 4000^2$

Earth

Note: not drawn to scale — you would be *way* smaller.

Solve It

17. Find all lines through (0, 1) normal to the curve $y = x^4$. The results may surprise you. Before you begin solving this, graph $y = x^4$ and put the cursor at (0, 1). Now guess where normal lines will be and whether they represent shortest paths or longest paths from (0, 1) to $y = x^4$. *Note:* Do *ZoomSqr* to get the distances on the graph to appear in their proper proportion.

Solve It

18. An ill-prepared adventurer has run out of water on a hot sunny day in the desert. He's 30 miles due north and 7 miles due east of his camp. His map shows a winding river — that by some odd coincidence happens to flow according to the function $y = 10\sin\frac{x}{10} + 10\cos\frac{x}{5} + x$ (where his camp lies at the origin). See the following figure. What point along the river is closest to him? He figures that he and his camel can just barely make it another 15 miles or so.

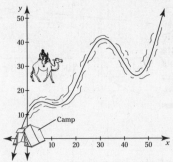

Solve It

Looking Smart with Linear Approximation

Linear approximation is easy to do, and once you get the hang of it, you can impress your friends by approximating things like $\sqrt[3]{70}$ in your head — like this: Bingo! 4.125. How did I do it? Look at Figure 8-2 and then at the example to see how I did it.

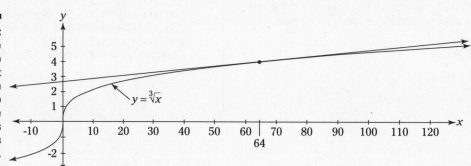

Figure 8-2: The line tangent to the curve at (64, 4) can be used to approximate cube roots of numbers near 64.

Q. Use linear approximation to estimate $\sqrt[3]{70}$.

A. 4.125

1. **Find a perfect cube root near $\sqrt[3]{70}$.**

   You notice that $\sqrt[3]{70}$ is near a no-brainer, $\sqrt[3]{64}$. That gives you the point (64, 4) on the graph of $y = \sqrt[3]{x}$.

2. **Find the slope of $y = \sqrt[3]{x}$ at $x = 64$.**

   $y' = \frac{1}{3} x^{-2/3}$ so the slope at 64 is $\frac{1}{48}$.

   This tells you that you add (or subtract) $\frac{1}{48}$ to 4 for each increase (decrease) of one from 64. For example, the cube root of 65 is $4\frac{1}{48}$ and the cube root of 66 is $4\frac{2}{48}$, or $4\frac{1}{24}$.

3. **Use the point-slope form to write the equation of the tangent line at (64, 4).**

   $$y - 4 = \frac{1}{48}(x - 64)$$
   $$y = \frac{1}{48}(x - 64) + 4$$

4. **Because this tangent line runs so close to the function $y = \sqrt[3]{x}$ near $x = 64$, use it to estimate cube roots of numbers near 64, namely at $x = 70$.**

   $$y = \frac{1}{48}(70 - 64) + 4$$
   $$= 4\frac{1}{8}$$

   By the way, in your calc text, the above simple point-slope form from algebra is probably rewritten in highfalutin calculus terms — like this:

   $$l(x) = f(x_0) + f'(x_0)(x - x_0)$$

   Don't be intimidated by this equation. It's just your friendly old algebra equation in disguise! Look carefully at it term by term and you'll see that it's mathematically identical to the point-slope equation tweaked like this:
   $$y = y_0 + m(x - x_0)$$

19. Estimate the 4th root of 17.

Solve It

20. Approximate 3.01^5.

Solve It

21. Estimate $\sin\frac{\pi}{180}$, that's one degree of course.

Solve It

22. Approximate $\ln(e^{10}+5)$.

Solve It

Solutions to Differentiation Problem Solving

1 What are the dimensions of the soup can of greatest volume that can be made with 50 square inches of tin? What's its volume? **The dimensions are 3¼ inches wide and 3¼ inches tall. The volume is 27.14 cubic inches.**

1. Draw your diagram (see the following figure).

2. a. Write a formula for the thing you want to maximize, the volume: $V = \pi r^2 h$

   b. Use the given information to relate r and h.

   $$Surface\ Area = \overbrace{2\pi r^2}^{top\ and\ bottom} + \overbrace{2\pi rh}^{lateral\ area}$$
   $$50 = 2\pi r^2 + 2\pi rh$$
   $$25 = \pi r^2 + \pi rh$$

   c. Solve for h and substitute to create a function of one variable.

   $$\pi rh = 25 - \pi r^2 \qquad\qquad V = \pi r^2 h$$
   $$h = \frac{25}{\pi r} - r \qquad\qquad V(r) = \pi r^2 \left(\frac{25}{\pi r} - r \right)$$
   $$= 25r - \pi r^3$$

3. Figure the domain.

   $r > 0$ is obvious

   $h > 0$ is also obvious

   And because $25 = \pi r^2 + \pi rh$ (from Step 2b), when $h = 0$, $r = \sqrt{\frac{25}{\pi}}$ so r must be less than $\sqrt{\frac{25}{\pi}}$, or about 2.82 inches.

4. Find the critical numbers of $V(r)$.

   $$V(r) = 25r - \pi r^3$$
   $$V'(r) = 25 - 3\pi r^2$$
   $$0 = 25 - 3\pi r^2$$
   $$r^2 = \frac{25}{3\pi}$$
   $$r = \pm\sqrt{\frac{25}{3\pi}}$$

   ≈ 1.63 inches $\big($You can reject the negative answer because it's outside the domain.$\big)$

5. Evaluate the volume at the critical number.

   $$V(1.63) = 25 \cdot 1.63 - \pi(1.63)^3$$
   $$\approx 27.14 \text{ cubic inches}$$

That's about 15 ounces. The can will be $2 \cdot 1.63$ or about 3¼ inches wide and $\dfrac{25}{\pi \cdot 1.63} - 1.63$ or about 3¼ inches tall. Isn't that nice? The largest can has the same width and height and would thus fit perfectly into a cube. Geometric optimization problems frequently have results where the dimensions have some nice, simple mathematical relation to each other.

2 ... What dimensions will minimize the cost of the frame? **The dimensions are 4'3" wide and 5'1" high. The minimum cost is $373.**

1. Draw a diagram with variables (see the following figure).

2. a. Express the thing you want to minimize, the cost.

$Cost = \left(length\ of\ curved\ frame\right) \cdot \left(cost\ per\ linear\ foot\right) +$

$\qquad \left(length\ of\ straight\ frame\right) \cdot \left(cost\ per\ linear\ foot\right)$

$\qquad = \left(\pi x\right)\left(25\right) + \left(2x + 2y\right)\left(20\right)$

$\qquad = 25\pi x + 40x + 40y$

b. Relate the two variables to each other.

$Area$ = Semicircle + Rectangle

$20 = \dfrac{\pi x^2}{2} + 2xy$

c. Solve for y and substitute.

$2xy = 20 - \dfrac{\pi x^2}{2}$ $\qquad cost = 25\pi x + 40x + 40y$

$y = \dfrac{20}{2x} - \dfrac{\pi x^2}{4x}$ $\qquad C\left(x\right) = 25\pi x + 40x + 40\left(\dfrac{10}{x} - \dfrac{\pi x}{4}\right)$

$\quad = \dfrac{10}{x} - \dfrac{\pi x}{4}$ $\qquad\qquad = 25\pi x + 40x + \dfrac{400}{x} - 10\pi x$

$\qquad\qquad\qquad\qquad = 15\pi x + 40x + \dfrac{400}{x}$

3. Find the domain.

$x > 0$ is obvious. And when x gets large enough, the entire window of 20 square feet in area will be one big semicircle, so

$20 = \dfrac{\pi x^2}{2}$

$40 = \pi x^2$

$x^2 = \dfrac{40}{\pi}$

$x = \sqrt{\dfrac{40}{\pi}}$

$\quad \approx 3.57$

Thus, x must be less than or equal to 3.57.

4. Find the critical numbers of $C(x)$.

$$C(x) = 15\pi x + 40x + \frac{400}{x}$$
$$C'(x) = 15\pi + 40 + (-400)x^{-2}$$
$$0 = 15\pi + 40 - 400x^{-2}$$
$$400x^{-2} = 15\pi + 40$$
$$x^2 = \frac{400}{15\pi + 40}$$
$$x = \pm\sqrt{\frac{400}{15\pi + 40}}$$
$$x \approx \pm 2.143$$

Omit –2.143 because it's outside the domain. So 2.143 is the only critical number.

5. Evaluate the cost at the critical number and at the endpoints.

$$C(x) = 15\pi x + 40x + \frac{400}{x}$$
$$C(0) = \text{undefined}$$
$$C(2.143) \approx \$373$$
$$C(3.57) \approx \$423$$

So, the least expensive frame for a 20-square-foot window will cost about \$373 and will be 2×2.143, or about 4.286 feet or 4'3" wide at the base. Because $y = \frac{10}{x} - \frac{\pi x}{4}$, the height of the rectangular lower part of the window will be 2.98, or about 3' tall. The total height will thus be 2.98 plus 2.14, or about 5'1".

3 . . . Given that a right triangle's hypotenuse must pass through the point (2, 5), what are the dimensions and area of the smallest such triangle? **The hypotenuse meets the y-axis at (0, 10) and the x-axis at (4, 0) and the triangle's area is 20.**

1. Draw a diagram (see the following figure).

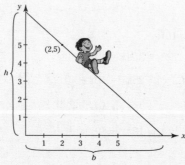

2. a. Write a formula for the thing you want to minimize, the area: $A = \frac{1}{2}bh$

b. Use the given constraints to relate b and h.

This is a bit tricky — *Hint:* consider similar triangles. If you draw a horizontal line from (0, 5) to (2, 5), you create a little triangle in the upper-left corner that's similar to the whole triangle. (You can prove their similarity with AA — remember your geometry? — both triangles have a right angle and both share the top angle.)

Because the triangles are similar, their sides are proportional:

$$\frac{height_{big\ triangle}}{base_{big\ triangle}} = \frac{height_{small\ triangle}}{base_{small\ triangle}}$$
$$\frac{h}{b} = \frac{h-5}{2}$$

c. Solve for one variable in terms of the other — take your pick — and substitute into your formula to create a function of a single variable.

$$2h = b(h-5)$$
$$2h = bh - 5b$$
$$h(2-b) = -5b$$
$$h = \frac{5b}{b-2}$$

$$A = \frac{1}{2}bh$$
$$A(b) = \frac{1}{2}b \cdot \left(\frac{5b}{b-2}\right)$$
$$= \frac{5b^2}{2b-4}$$

3. Find the domain.

b must be greater than 2 — do you see why? And there's no maximum value for b.

4. Find the critical numbers.

$$A(b) = \frac{5b^2}{2b-4}$$
$$A'(b) = \frac{(5b^2)'(2b-4) - (5b^2)(2b-4)'}{(2b-4)^2}$$
$$= \frac{10b(2b-4) - 10b^2}{(2b-4)^2}$$
$$= \frac{10b^2 - 40b}{(2b-4)^2}$$

$$\frac{10b^2 - 40b}{(2b-4)^2} = 0$$
$$10b^2 - 40b = 0$$
$$10b(b-4) = 0$$
$$b = 0 \ \text{ or } \ 4$$

Zero is outside the domain, so 4 is the only critical number. The smallest triangle must occur at $b = 4$ because near the endpoints you get triangles with astronomical areas.

5. Finish.

$$b = 4$$
$$h = \frac{5b}{b-2} \quad \text{so}$$
$$h = \frac{5 \cdot 4}{4-2} = 10;$$

And the triangle's area is thus 20.

4 ... Given that you want a box with a volume of 72 cubic inches, what dimensions will minimize the total cardboard area and thus minimize the cost of the cardboard? **The minimizing dimensions are 6-by-6-by-2, made with 108 square inches of cardboard.**

1. Draw a diagram and label with variables (see the following figure).

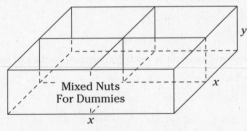

2. a. Express the thing you want to minimize, the cardboard area, as a function of the variables.

$$Cardboard\ area = \overbrace{x^2}^{square\ base} + \overbrace{4xy}^{four\ sides} + \overbrace{2xy}^{two\ dividers}$$
$$A = x^2 + 6xy$$

b. Use the given constraint to relate x to y.

$$Vol = l \cdot w \cdot h$$
$$72 = x \cdot x \cdot y$$

c. Solve for y and substitute in equation from Step 2a to create a function of one variable.

$$y = \frac{72}{x^2}$$
$$A = x^2 + 6xy$$
$$A(x) = x^2 + 6x\left(\frac{72}{x^2}\right)$$
$$= x^2 + \frac{432}{x}$$

3. Find the domain.

$x > 0$ is obvious

$y > 0$ is also obvious

And if you make y small enough, say the height of a proton — great box, eh? — x would have to be astronomically big to make the volume 72 cubic inches. Technically, there is no maximum value for x.

4. Find the critical numbers.

$$A(x) = x^2 + \frac{432}{x}$$
$$A'(x) = 2x - 432x^{-2}$$
$$0 = 2x - \frac{432}{x^2}$$
$$\frac{432}{x^2} = 2x$$
$$x = \sqrt[3]{216}$$
$$= 6$$

You know this number has to be a minimum because near the endpoints, say when $x = .0001$ or $y = .0001$, you get absurd boxes — either thin and tall like a mile-high toothpick or short and flat like a square piece of cardboard as big as a city block with a microscopic lip. Both of these would have *enormous* area and would be of interest only to calculus professors.

5. Finish.

$x = 6$, so the total area is

$$A(6) = 6^2 + \frac{432}{6} \qquad \text{Because } y = \frac{72}{x^2}$$
$$= 36 + 72 \qquad\qquad\quad y = 2$$
$$= 108$$

That's it — a 6-by-6-by-2 box made with 108 square inches of cardboard.

5 ... When the depth of the swill falls to 1 foot 8 inches, how fast is the swill level falling? **It's falling at a rate of ⁹⁄₁₀ inches per minute.**

1. Draw a diagram, labeling the diagram with any *unchanging* measurements and assigning variables to any *changing* things. See the following figure.

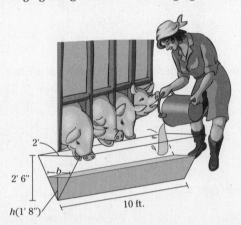

Note that the figure shows the *unchanging* dimensions of the trough, 2 feet by 2 feet 6 inches by 10 feet, and these dimensions are *not* labeled with variable names like *l* (for length), *w* (for width), or *h* (for height). Also note that the *changing* things — the height (or depth) of the swill and the width of the surface of the swill (which gets narrower as the swill level falls) — *do* have variable names, *h* for height and *b* for base (I realize it's at the top, but it's the base of the upside-down triangle shape made by the swill). Finally note that the height of 1'8" — which is the height only at one particular point in time — is in parentheses to distinguish it from the other *unchanging* dimensions.

2. List all given rates and the rate you're asked to figure out. Express these rates as derivatives with respect to time. Give yourself a high-five if you realized that the thing that matters about the changing volume of swill is the *net* rate of change of volume.

Swill is coming in at 1 *cubic foot per minute* and is going out at $3 \cdot \frac{1}{2}$ *cubic feet per minute* (for the three hogs) plus another ½ *cubic feet per minute* (the splashing). So the net is 1 *cubic foot per minute* going out — that's a *negative* rate of change. In calculus language, you write:

$$\frac{dv}{dt} = -1 \text{ cubic foot per minute}.$$

You're asked to determine how fast the height is changing, so write:

$$\frac{dh}{dt} = ?$$

3. a. Write down a formula that involves the variables in the problem — *V*, *h*, and *b*.

The technical name for the shape of the trough is a *right prism*. And the shape of the swill in the trough — what you care about here — has the same shape. Imagine tipping this up so it stands vertically. Any shape that has a flat base and a flat top and that goes straight up from base to top has the same volume formula: *Volume = area_{base} · height*

Note that this "base" is the entire swill triangle and totally different from *b* in the figure; also this "height" is totally different from the swill height, *h*.

The area of the triangular base equals $\frac{1}{2}bh$ and the height of the prism is 10 feet, so here's your formula: $V = \frac{1}{2}bh \cdot 10 = 5bh$

Because *b* doesn't appear in your list of derivatives in Step 2, you want to get rid of it.

b. Find an equation that relates your unwanted variable, *b*, to some other variable in the problem so you can make a substitution and be left with an equation involving only *V* and *h*. The triangular face of the swill is the same shape of the triangular side of the trough. If you remember geometry, you know that such similar shapes have proportional sides. So,

$$\frac{b}{2} = \frac{h}{2.5}$$
$$2.5b = 2h$$
$$b = .8h$$

Similar triangles often come up in related rate problems involving triangles, triangular prisms, and cones.

Now substitute .8*h* for *b* in the formula from Step 3a:

$$V = 5bh$$
$$= 5 \cdot .8h \cdot h$$
$$= 4h^2$$

4. Differentiate with respect to *t*.

$$\frac{dv}{dt} = 8h\frac{dh}{dt}$$

In all related rates problems, make sure you differentiate (like you do here in Step 4) *before* you substitute the values of the variables into the equation (like you do below when you plug 1'8" into *h* in Step 5).

5. Substitute all known quantities into this equation and solve for $\frac{dh}{dt}$.

You were given that *h* = 1'8" (you must convert this to feet) and you figured out in Step 2 that $\frac{dv}{dt} = -1$, so

$$-1 = 8 \cdot 1\frac{2}{3} \cdot \frac{dh}{dt}$$
$$\frac{dh}{dt} = \frac{-1}{\frac{40}{3}}$$
$$= \frac{-3}{40} \, ft/min$$
$$= \frac{-9}{10} \, inches/min$$

Thus, when the swill level drops to a depth of 1'8", it's falling at a rate of 9⁄10 inches per minute. Mmm, mmm, good!

6. Ask whether this answer makes sense.

Unlike the example problem, it's not easy to come up with a common-sense explanation of why this answer is or is not reasonable. But there's another type of check that works here and in many other related rates problems.

Take a very small increment of time — something much less than the time unit of the rates used in the problem. This problem involves rates per *minute,* so use 1 second for your time increment. Now ask yourself what happens in this problem in 1 second. The swill is leaving the trough at 1 *cubic foot / minute;* so in 1 second, 1⁄60 cubic foot will leave the trough. What does that do to the swill height? Because of the similar triangles mentioned in Step 3b, when the swill falls to a depth of 1'8", which is ⅔ of the height of the trough, the width of the surface of the swill must be ⅔ of the width of the trough — and that comes to 1⅓ feet. So the surface area of the swill is $1\frac{1}{3} \times 10$ feet.

Assuming the trough walls are straight (this type of simplification always works in this type of checking process), the swill that leaves the trough would form the shape of a *very, very* short box ("box" sounds funny because this shape is so thin; maybe "thin piece of plywood" is a better image).

The volume of a box equals *length · width · height*, thus

$$\frac{1}{60} = 10 \cdot 1\frac{1}{3} \cdot height$$
$$height = .00125$$

This tells you that in 1 second, the height should fall .00125 feet or something very close to it. (This process sometimes produces an exact answer and sometimes an answer with a very small error.) Now, finally, see whether this number agrees with the answer. Your answer was –⁹⁄₁₀ *inches/minute*. Convert this to *feet/second*:

$$-\frac{9}{10} \div 12 \div 60 = -.00125. \text{ It checks.}$$

6 ... When it reaches a height of 60 feet, it's moving up at a rate of 50 *ft/sec*. At this point, how fast is the distance from 2nd base to the ball growing? **The distance is growing 21.3 feet/second.**

1. Draw your diagram and label it. See the following figure.

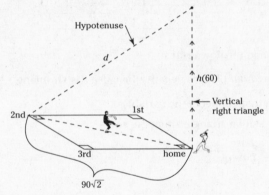

2. List all given rates and the rate you're asked to figure out.

$$\frac{dh}{dt} = 50 \ ft/sec$$

$$\frac{dd}{dt} = ?$$

3. Write a formula that involves the variables: $h^2 + \left(90\sqrt{2}\right)^2 = d^2$

4. Differentiate with respect to time: $2h\frac{dh}{dt} = 2d\frac{dd}{dt}$

Like in the example, you're missing a needed value, *d*. So use the Pythagorean Theorem to get it:

$$h^2 + \left(90\sqrt{2}\right)^2 = d^2$$
$$60^2 + \left(90\sqrt{2}\right)^2 = d^2$$
$$d \approx \pm 140.7 \text{ feet} \ \left(\text{You can reject the negative answer.}\right)$$

Now do the substitutions:

$$2h\frac{dh}{dt} = 2d\frac{dd}{dt}$$
$$2 \cdot 60 \cdot 50 = 2 \cdot 140.7\frac{dd}{dt}$$
$$\frac{dd}{dt} = \frac{2 \cdot 60 \cdot 50}{2 \cdot 140.7}$$
$$\approx 21.3 \ ft/sec$$

5. Check whether this answer makes sense.

For this one, you're on your own. *Hint:* Use the Pythagorean Theorem to calculate d $\frac{1}{50}$ second after the critical moment. Do you see why I picked this time increment?

7 ... Five feet before the man crashes into the lamp post, he's running at a speed of 15 *miles/hour*. At this point, how fast is the tip of the shadow moving? **It's moving at 25 *miles/hour*.**

1. The diagram thing: See the following figure.

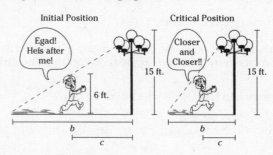

2. List the known and unknown rates.

$$\frac{dc}{dt} = -15 \ miles/hour \ \text{(This is negative because c is shrinking.)} \quad \frac{db}{dt} = ?$$

3. Write a formula that connects the variables.

This is another similar triangle situation, so —

$$\frac{height_{\text{big triangle}}}{height_{\text{little triangle}}} = \frac{base_{\text{big triangle}}}{base_{\text{little triangle}}}$$

$$\frac{15}{6} = \frac{b}{b-c}$$

$$15b - 15c = 6b$$

$$9b = 15c$$

$$3b = 5c$$

4. Differentiate with respect to t: $3\dfrac{db}{dt} = 5\dfrac{dc}{dt}$

5. Substitute known values.

$$3\frac{db}{dt} = 5(-15)$$

$$\frac{db}{dt} = -25 \ miles/hour$$

Thus, the top of the shadow is moving toward the lamp post at 25 *miles/hour* — and is thus gaining on the man at a rate of 10 *miles/hour*.

A somewhat unusual twist in this problem is that you never had to plug in the given distance of 5 ft. This is because the speed of the shadow is independent of the man's position.

8 ... If the height of the cone-shaped pile is always equal to the radius of the cone's base, how fast is the height of the pile increasing when it's 18 feet tall? **It's increasing at 2⅛ *inches/min*.**

1. Draw your diagram: See the following figure.

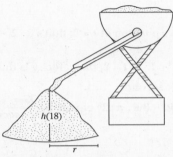

2. List the rates: $\dfrac{dv}{dt} = 200$ *cubic ft/min* $\dfrac{dh}{dt} = ?$

3. a. The formula thing: $V_{cone} = \dfrac{1}{3}\pi r^2 h$

   b. Write an equation relating r and h so that you can get rid of r: $r = h$

   What could be simpler? Now get rid of r: $V = \dfrac{1}{3}\pi h^2 h = \dfrac{1}{3}\pi h^3$

4. Differentiate: $\dfrac{dV}{dt} = \pi h^2 \dfrac{dh}{dt}$

5. Substitute and solve for $\dfrac{dh}{dt}$.

   $200 = \pi \cdot 18^2 \dfrac{dh}{dt}$

   $\dfrac{dh}{dt} \approx .196$ *ft/min* $\approx 2\dfrac{1}{3}$ *inches/min*

6. Check whether this answer makes sense.

   Calculate the increase in the height of the cone from the critical moment ($h = 18$) to ⅟₂₀₀ minute after the critical moment. When $h = 18$, $V = \dfrac{1}{3}\pi(18)^3$, or about 6107.256 cubic feet. ⅟₂₀₀ minute later, the volume (which grows at a rate of 200 *cubic feet per minute*) will increase by 1 cubic foot to about 6108.256 cubic feet. Now solve for h:

   $6108.256 = \dfrac{1}{3}\pi h^3$

   $h = \sqrt[3]{\dfrac{6108.256}{\dfrac{1}{3}\pi}}$

   ≈ 18.000982

   Thus, in ⅟₂₀₀ minute, the height would grow from 18 feet to 18.000982 feet. That's a change of .000982 feet. Multiply that by 200 to get the change in 1 minute: $.000982 \cdot 200 \approx .196$

   It checks.

9 $s(t) = 5t^2 + 4$

a. At $t = 2$, the platypus's position is $s(t) = 24$ feet from the back of your boat.

b. $v(t) = s'(t) = 10t$, so at $t = 2$, the platypus's velocity is $s'(2) = 20$ *feet/second* (20 is positive so that's toward the front of the boat).

c. Speed is the absolute value of velocity, so the speed is also 20 ft/sec.

d. Acceleration, $a(t)$, equals $v'(t) = s''(t) = 10$. That's a constant, so the platypus's acceleration is $10\dfrac{feet/second}{second}$ at all times.

10 $s(t) = 3t^4 - 5t^3 + t - 6$

    a. $s(2)$ gives the platypus's position at $t = 2$; that's $3 \cdot 2^4 - 5 \cdot 2^3 + 2 - 6$, or 4 feet, from the back of the boat.

    b. $v(t) = s'(t) = 12t^3 - 15t^2 + 1$. At $t = 2$, the velocity is thus 37 *feet per second*.

    c. Speed is also 37 feet per second.

    d. $a(t) = v'(t) = s''(t) = 36t^2 - 30t$. $a(2)$ equals 84 $\dfrac{feet/second}{second}$.

11 $s(t) = \dfrac{1}{t} + \dfrac{8}{t^3} - 3$

    a. At $t = 2$, $s(2)$ equals $\dfrac{1}{2} + 1 - 3$, or $-1\frac{1}{2}$ feet. This means that the platypus is 1½ feet *behind* the back of the boat.

    b. $v(t) = s'(t) = -t^{-2} + \left(-24 t^{-4}\right)$

$$v(2) = s'(2) = -2^{-2} + \left(-24(2)^{-4}\right)$$

$$= -\frac{1}{4} - \frac{24}{16}$$

$$= -1\frac{3}{4} \ feet/second$$

A negative velocity means that the platypus is swimming "backwards," in other words away from the back of the boat.

    c. *Speed* $= |Velocity|$, so the platypus's speed is $1\frac{3}{4}$ *feet/second.*

    d. $a(t) = v'(t) = s''(t) = 2t^{-3} + 96t^{-5}$ or $\dfrac{2}{t^3} + \dfrac{96}{t^5}$. $a(2)$ is therefore $\dfrac{2}{8} + \dfrac{96}{32}$, or $3\frac{1}{4} \ \dfrac{feet/second}{second}$.

Give yourself a pat on the back if you figured out that this positive acceleration with a negative velocity means the platypus is actually slowing down.

12 $s(t) = 2t^3 - t^2 + 8t - 5$

    a. Find the zeros of the velocity:

$$v(t) = s'(t) = 6t^2 - 2t + 8$$
$$0 = 6t^2 - 2t + 8$$
$$= 3t^2 - t + 4$$

No solutions because the discriminant is negative.

The discriminant equals $b^2 - 4ac$.

The fact that the velocity is never zero means that the sloth never turns around. At $t = 0$, $v(t) = 8 \ ft/sec$ which is positive, so the sloth moves away from the trunk for the entire interval $t = 0$ to $t = 5$.

    b. and c. Because there are no turnaround points and because the motion is in the positive direction, the total distance and total displacement are the same: 265 feet.

$$s(5) - s(0) = 260 - (-5) = 265$$

Whenever the total displacement equals the total distance, average velocity also equals average speed: 53 *ft/sec.*

$$\frac{total \ displacement}{total \ time} = \frac{s(5) - s(0)}{5 - 0} = \frac{265}{5} = 53 \ ft/sec.$$

13 $s(t) = t^4 + t^2 - t$

a. Find the zeros of $v(t)$: $v(t) = s'(t) = 4t^3 + 2t - 1$

You'll need your calculator for this:

Graph $y = 4x^3 + 2x - 1$ and locate the x-intercepts. There's just one: $x \approx .385$. That's the only zero of $s'(t) = v(t)$.

Don't forget that a zero of a derivative can be a horizontal inflection as well as a local extremum. You get a turnaround point only at the local extrema.

Because $v(0) = -1$ (a leftward velocity) and $v(1) = 5$ (a rightward velocity), $s(.385)$ must be a turnaround point (and it's also a local min on the position graph). Does the first derivative test ring a bell?

Thus, the sloth is going left **from $t = 0$ sec to $t = .385$ sec and right from .385 to 5 sec. He turns around, obviously, at .385 sec when he is at $s(.385) = .385^4 + .385^2 - .385$ or $-.215$ meters — that's .215 meters to the left of the trunk.** I presume you figured out that there must be another branch on the tree on the other side of the trunk to allow the sloth to go left to a negative position.

b. There are two legs of the sloth's trip. He goes left from $t = 0$ till $t = .385$, then right from $t = .385$ till $t = 5$. Just add up the *positive* lengths of the two legs.

$$length_{\text{leg 1}} = \left| s(.385) - s(0) \right|$$
$$= \left| -.215 - 0 \right|$$
$$= .215 \text{ meters}$$

$$length_{\text{leg 2}} = \left| s(5) - s(.385) \right|$$
$$= \left| 5^4 + 5^2 - 5 - (-.215) \right|$$
$$= 645.215 \text{ meters}$$

The total distance is thus .215 + 645.215, or 645.43 meters. That's one big tree! The branch is over 2000 feet long.

His average speed is 645.43 / 5, or about 129.1 *meters/second*. That's one fast sloth! Almost 300 *miles/hour*!

c. **Total displacement is $s(5) - s(0)$, that's $645 - 0 = 645$ meters. Lastly, his average velocity is simply total displacement divided by total time — that's 645/5, or 129 *meters/second*.**

14 $s(t) = \dfrac{t+1}{t^2+4}$

a. Find the zeros of $v(t)$:

$$s'(t) = v(t) = \frac{(t+1)'(t^2+4) - (t+1)(t^2+4)'}{(t^2+4)^2}$$
$$= \frac{t^2 + 4 - (2t^2 + 2t)}{(t^2+4)^2}$$
$$= \frac{-t^2 - 2t + 4}{(t^2+4)^2}$$

Set this equal to zero and solve:

$$\frac{-t^2 - 2t + 4}{(t^2+4)^2} = 0$$
$$t^2 + 2t - 4 = 0$$
$$t = \frac{-2 \pm \sqrt{4 - (-16)}}{2}$$
$$\approx -3.236 \text{ or } 1.236$$

Reject the negative solution because it's outside the interval of interest: $t = 0$ to $t = 5$. So, the only zero velocity occurs at $t = 1.236$ seconds.

Because $v(0) = .25$ *meters/second* and $v(5) \approx -.037$, the first derivative test tells you that $s(1.236)$ must be a local max and therefore a turnaround point.

The sloth thus goes right from $t = 0$ till $t = 1.236$ seconds, then turns around at $s(1.236)$, or about .405 meters to the right of the trunk, and goes left till $t = 5$.

b. His total distance is the sum of the lengths of the two legs:

$$\text{going right} = \left| s(1.236) - s(0) \right|$$
$$= |.405 - .25|$$
$$\approx .155$$

$$\text{going left} = \left| s(5) - s(1.236) \right|$$
$$\approx .198$$

Total distance is therefore .155 + .198 = .353 meters. His average speed is thus .353/5, or .071 *meters/second*. That's roughly a sixth of a *mile/hour* — *much* more like it for a sloth. That's pretty darn slow, but how quick do you think you'd be with only three toes?

c. Total displacement is defined as final position minus initial position, so that's

$$s(5) - s(0) = \frac{6}{29} - \frac{1}{4}$$
$$\approx -.043 \text{ meters}$$

And thus his average velocity is $-.043/5$, or $-.0086$ *meters/second*. You're done.

15 Two lines through the point $(1, -3)$ are tangent to the parabola $y = x^2$. Determine the points of tangency. **The points of tangency are $(-1, 1)$ and $(3, 9)$.**

1. Express a point on the parabola in terms of x.

The equation of the parabola is $y = x^2$, so you can take a general point on the parabola (x, y) and substitute x^2 for y. So your point is (x, x^2).

2. Take the derivative of the parabola.

$$y = x^2$$
$$y' = 2x$$

3. Using the slope formula, $m = \frac{y_2 - y_1}{x_2 - x_1}$, set the slope of the tangent line from $(1, -3)$ to (x, x^2) equal to the derivative. Then solve for x.

$$\frac{x^2 - (-3)}{x - 1} = 2x$$
$$x^2 + 3 = 2x^2 - 2x$$
$$x^2 - 2x - 3 = 0$$
$$(x + 1)(x - 3) = 0$$
$$x = -1 \text{ or } 3$$

4. Plug these x-coordinates into $y = x^2$ to get the y-coordinates.

$$y = (-1)^2 = 1$$
$$y = 3^2 = 9$$

So there's one line through $(1, -3)$ that's tangent to the parabola at $(-1, 1)$ and another through $(1, -3)$ that's tangent at $(3, 9)$. You may want to confirm these answers by graphing the parabola and your two tangent lines:

$$y = -2(x + 1) + 1$$
$$y = 6(x - 3) + 9$$

16 ... How far out can you see to the horizon before the Earth's curvature makes the water dip below the horizon? **You can see out 2.83 miles.**

1. Write the equation of the Earth's circumference as a function of y (see the figure in the problem).

$$x^2 + y^2 = 4000^2$$
$$y = \pm\sqrt{4000^2 - x^2}$$

You can disregard the negative half of this circle because your line of sight will obviously be tangent to the upper half of the Earth.

2. Express a point on the circle in terms of x: $\left(x, \sqrt{4000^2 - x^2}\right)$.

3. Take the derivative of the circle.

$$y = \sqrt{4000^2 - x^2}$$
$$y' = \frac{1}{2}\left(4000^2 - x^2\right)^{-1/2}(-2x) \quad \text{(Chain Rule)}$$
$$= \frac{-x}{\sqrt{4000^2 - x^2}}$$

4. Using the slope formula, set the slope of the tangent line from your eyes to $\left(x, \sqrt{4000^2 - x^2}\right)$ equal to the derivative and then solve for x.

Your eyes are 5' 3.36" above the top of the Earth at the point (0, 4000) on the circle. Convert your height to miles, that's exactly .001 miles (What an amazing coincidence!). So the coordinates of your eyes are (0, 4000.001).

$$\frac{y_2 - y_1}{x_2 - x_1} = m$$
$$\frac{\sqrt{4000^2 - x^2} - 4000.001}{x - 0} = \frac{-x}{\sqrt{4000^2 - x^2}}$$
$$-x^2 = \left(4000^2 - x^2\right) - 4000.001\sqrt{4000^2 - x^2} \quad \text{(Cross multiplication)}$$
$$-4000^2 = -4000.001\sqrt{4000^2 - x^2} \quad \text{(Use your calculator, of course)}$$
$$3999.999 = \sqrt{4000^2 - x^2} \quad \text{(Now square both sides)}$$
$$15999992 = 4000^2 - x^2$$
$$x^2 = 8$$
$$x = 2\sqrt{2} \approx 2.83 \text{ miles}$$

Many people are surprised that the horizon is so close. What do you think?

17 Find all lines through (0, 1) normal to the curve $y = x^4$. **Five normal lines can be drawn to $y = x^4$ from (0, 1). The points of normalcy are (–.915, .702), (–.519, .073), (0, 0), (.519, .073), and (.915, .702).**

1. Express a point on the curve in terms of x: A general point is $\left(x, x^4\right)$.

2. Take the derivative.

$$y = x^4$$
$$y' = 4x^3$$

3. Set the slope from $(0, 1)$ to $\left(x, x^4\right)$ equal to the opposite reciprocal of the derivative and solve.

$$\frac{x^4 - 1}{x - 0} = \frac{-1}{4x^3}$$
$$4x^7 - 4x^3 + x = 0$$
$$x\left(4x^6 - 4x^2 + 1\right) = 0 \qquad x = 0 \quad \text{or} \quad 4x^6 - 4x^2 + 1 = 0$$

Unless you have a special gift for solving 6th degree equations, you better use your calculator — just graph $y = 4x^6 - 4x^2 + 1$ and find all of the x-intercepts. There are x-intercepts at ~–.915, ~–.519, ~.519, and ~.915. Dig those palindromic numbers!

4. Plug these four solutions into $y = x^4$ to get the y-coordinates. And there's also the $x = 0$ no-brainer.

$$(-.519)^4 = (.519)^4 = \sim.073$$
$$(-.915)^4 = (.915)^4 = \sim.702$$

You're done. Five normal lines can be drawn to $y = x^4$ from (0, 1). The points of normalcy are (–.915, .702), (–.519, .073), (0, 0), (.519, .073), and (.915, .702).

I find this result interesting. First, because there are so many normal lines, and second, because the normal lines from (0, 1) to (–.915, .702), (0, 0), and (.915, .702) are all *shortest* paths (compared to other points in their respective vicinities). The other two normals are longest paths. This is curious because $y = x^4$ is everywhere concave *toward* (0, 1). When a curve is concave away from a point, a normal to the curve can only be a local shortest path. But when a curve is concave toward a point, you can get either a local shortest or a local longest path.

I played slightly fast and loose with the math for the $x = 0$ solution. Did you notice that $x = 0$ doesn't work if you plug it back into the equation, $\frac{x^4 - 1}{x - 0} = \frac{-1}{4x^3}$, because both denominators become zero? However — promise not to leak this to your calculus teacher — this is okay here because both sides of the equation become $\frac{Non-zero\ number}{zero}$. (Actually, they're both $\frac{-1}{0}$, but something like $\frac{5}{0} = \frac{2}{0}$ would also work.) Non-zero over zero means a vertical line with undefined slope. So the $\frac{-1}{0} = \frac{-1}{0}$ tells you that you've got a vertical normal line at $x = 0$.

18 ... What point along the river is closest to the adventurer? **The closest point is (6.11, 15.26), which is 14.77 miles away.**

1. Express a point on the curve in terms of x: $\left(x,\ 10\sin\frac{x}{10} + 10\cos\frac{x}{5} + x \right)$

2. Take the derivative.

$$y = 10\sin\frac{x}{10} + 10\cos\frac{x}{5} + x$$
$$y' = 10\cos\left(\frac{x}{10}\right) \cdot \frac{1}{10} - 10\sin\left(\frac{x}{5}\right) \cdot \frac{1}{5} + 1$$
$$= \cos\frac{x}{10} - 2\sin\frac{x}{5} + 1$$

3. Set the slope from (7, 30) to the general point equal to the opposite reciprocal of the derivative and solve.

$$\frac{30 - \left(10\sin\frac{x}{10} + 10\cos\frac{x}{5} + x\right)}{7 - x} = \frac{-1}{\cos\frac{x}{10} - 2\sin\frac{x}{5} + 1}$$

Unless you wear a pocket protector, don't even think about solving this equation without a calculator.

Solve on your calculator by graphing the following equation and finding the x-intercepts —

$$y = \frac{30 - \left(10\sin\frac{x}{10} + 10\cos\frac{x}{5} + x\right)}{7 - x} - \frac{-1}{\cos\frac{x}{10} - 2\sin\frac{x}{5} + 1}$$

It's a bit tricky to find the *x*-intercepts for this hairy function. You have to play around with the *window* settings a bit. And don't forget that your calculator will draw vertical asymptotes that look like zeros of the function, but are not. Now, it turns out that this function has an infinite number of x-intercepts (I think). There's one between x= –18 and –19 and more at bigger negatives. And there's one between x= 97 and 98 and more at bigger positives. But these zeros represent points on te river so far away that they neednot be considered. Only three zeros are plausible candidates for the closest trip to the river. To see the first zero, set *xmin* = –1, *xmax* = 10, *xscl* = 1, *ymin* = –5, *ymax* = 25, and *yscl* = 5. To see the other two, set *xmin* = 10, *xmax* = 30, *xscl* = 1, *ymin* = –2, *ymax* = 10, and *yscl* = 1. These zeros are at roughly 6.11, 13.75, and 20.58.

4. Plug the zeros into the original function to obtain the *y*-coordinates. You get the following points of normalcy: (6.11, 15.26), (13.75, 14.32), (20.58, 23.80).

5. Use the distance formula, $D = \sqrt{(x_2 - x_1)^2 + (y_2 - y_1)^2}$, to find the distance from our parched adventurer to the three points of normalcy.

   The distances are 14.77 miles to (6.11, 15.26), 17.07 miles to (13.75, 14.32), and 14.93 miles to (20.58, 23.80). Using his trusty compass, he heads mostly south and a little west to (6.11, 15.26). An added benefit of this route is that it's in the direction of his camp.

19 Estimate the 4th root of 17. **The approximation is 2.03125.**

1. Write a function based on the thing you're trying to estimate: $f(x) = \sqrt[4]{x}$

2. Find a "round" number near 17 where the 4th root is very easy to get: that's 16, of course. And you know $\sqrt[4]{16} = 2$. So (16, 2) is on *f*.

3. Determine the slope at your point.

$$f(x) = \sqrt[4]{x}$$
$$f'(x) = \frac{1}{4}x^{-3/4}$$
$$f'(16) = \frac{1}{32}$$

4. Use the point-slope form of a line to write the equation of the tangent line at (16, 2).

$$y - 2 = \frac{1}{32}(x - 16)$$

5. Plug your number into the tangent line and you've got your approximation.

$$y = \frac{1}{32}(17 - 16) + 2$$
$$= 2\frac{1}{32} \text{ or } 2.03125$$

The exact answer is about 2.03054. Your estimate is only $\frac{7}{100}$ of 1 percent too big! Not too shabby. Extra credit question (solve this or we may have to vote you off the island): No matter what 4th root you estimate with linear approximation, your answer will be too big. Do you see why?

20 Approximate 3.01^5. **The approximation is 247.05.**

1. Write your function: $g(x) = x^5$

2. Find your round number. That's 3, well duhh. So your point is (3, 243).

3. Find the slope at your point.

$$g'(x) = 5x^4$$
$$g'(3) = 405$$

4. Tangent line equation.

$$y - y_1 = m(x - x_1)$$
$$y - 243 = 405(x - 3)$$

5. Get your approximation: $y = 405(3.01 - 3) + 243 = 247.05$

   Only $\frac{1}{100}$ of a percent off.

21 Estimate $\sin\frac{\pi}{180}$, that's one degree of course. **The approximation is $\frac{\pi}{180}$.**

You know the routine

$$p(x) = \sin x \qquad\qquad y - y_1 = m(x - x_1)$$
$$p(0) = 0 \quad\rightarrow (0, 0) \text{ is the point} \qquad y - 0 = 1(x - 0)$$
$$p'(x) = \cos x \qquad\qquad y = x$$
$$p'(0) = 1 \quad\rightarrow 1 \text{ is the slope}$$

Your number is $\frac{\pi}{180}$, so you get $y = \frac{\pi}{180}$.

Which shows that for very small angles, the sine of the angle and the angle itself are approximately equal. (The same is true of the tangent of an angle, by the way.) $\frac{\pi}{180}$ is only ⅟₂₀₀% too big!

22 Approximate $\ln(e^{10} + 5)$. **The approximation is $10 + \frac{5}{e^{10}}$.**

Just imagine all the situations where such an approximation will come in handy!

$$q(x) = \ln(x) \qquad\qquad y - y_1 = m(x - x_1)$$
$$q(e^{10}) = 10 \quad\rightarrow (e^{10}, 10) \text{ is the point} \qquad y - 10 = \frac{1}{e^{10}}(x - e^{10}) \quad\rightarrow \text{ the tangent line}$$
$$q'(x) = \frac{1}{x} \qquad\qquad y = \frac{1}{e^{10}}\left((e^{10} + 5) - e^{10}\right) + 10$$
$$q'(e^{10}) = \frac{1}{e^{10}} \quad\rightarrow \frac{1}{e^{10}} \text{ is the slope} \qquad = 10 + \frac{5}{e^{10}}$$

Hold on to your hat. This answer is a mere 0.00000026% too big.

Part IV
Integration and Infinite Series

The 5th Wave   By Rich Tennant

CHALK ...
ERASER ...
CALCULATOR ...

Stanley's
Math Parlour

"It's 'Fast Herschel Fenniman,' the most nortorious math hustler of all time. If he asks if you'd like to run some trigonometric integrals with him, just walk away."

In this part . . .

*I*ntegration, like differentiation, is a highfalutin calculus word for a simple idea: addition. Every integration problem involves addition in one way or another. What makes integration such a powerful tool is that it sort of cuts up something (the weight of a large dome, the length of a power cable, the pressure on the walls of a pipe, and so on) into infinitely small chunks and then adds up the infinite number of chunks to arrive at a precise total. Without integration, many such problems can only be approximated. Part IV gives you practice with integration basics, techniques for finding integrals, and problem solving with integration.

Infinite series is a topic full of bizarre, counter-intuitive results that have fascinated thinkers for over 2,000 years. Zeno of Elea (5th century B.C.) gave us his famous paradox about the race between Achilles and the tortoise (the resolution of the paradox involves limits — see *Calculus For Dummies*). Your task with infinite series problems is to decide whether the sum of an infinitely long list of numbers *diverges* (adds up to infinity — usually) or *converges* (adds up to an ordinary, finite number).

Chapter 9

Getting into Integration

• •

In This Chapter

▶ Reconnoitering rectangles

▶ Trying trapezoids

▶ Summing sigma sums

▶ Defining definite integration

• •

*I*n this chapter, you begin the second major topic in calculus: integration. With integration you can find the total area or volume of weird shapes that, unlike triangles, spheres, cones, and other basic shapes, don't have simple area or volume formulas. You can use integration to total up other things as well. The basic idea is that — with the magic of limits — the thing you want the total of is cut up into infinitesimal pieces and then the infinite number of pieces are added up. But before moving on to integration, you warm up with some easy stuff: pre-pre-pre-calc — the area of rectangles.

By the way, despite the "kid stuff" quip, much of the material in this chapter and the first section of Chapter 10 is both more difficult and less useful than what follows it. If ever there was a time for the perennial complaint — "What is the point of learning this stuff?" — this is it. Now, some calculus teachers would give you all sorts of fancy arguments and pedagogical justifications for why this material is taught, but, let's be honest, the sole purpose of teaching these topics is to inflict maximum pain on calculus students. Well, you're stuck with it, so deal with it. The good news is that this material will make everything that comes later seem easy by comparison.

Adding Up the Area of Rectangles: Kid Stuff

The material in this section — using rectangles to approximate the area of strange shapes — is part of every calculus course because integration rests on this foundation. But, in a sense, this material doesn't involve calculus at all. You could do everything in this section without calculus, and if calculus had never been invented, you could still approximate area with the methods described here.

Q. Using 10 right rectangles, estimate the area under $f(x) = \ln x$ from 1 to 6.

A. **The approximate area is 6.181.**

1. **Sketch $f(x) = \ln x$ and divide the interval from 1 to 6 into ten equal increments. Each increment will have a length of ½ of course. See the figure in Step 2.**

2. **Draw a *right* rectangle for each of the ten increments.**

   You're doing *right* rectangles, so put your pen on the *right* end of the base of the first rectangle (that's at $x = 1.5$), draw straight up till you hit the curve, and then straight left till you're directly above the left end of the base ($x = 1$). Finally, going straight down, draw the left side of the first rectangle. See the following figure. I've indicated with arrows how you draw the first rectangle. Draw the rest the same way.

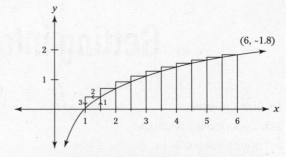

3. **Use your calculator to calculate the heights of each rectangle.**

   The heights are given by $f(1.5)$, $f(2)$, $f(2.5)$, and so on, which are $\ln 1.5$, $\ln 2$, and so on again.

4. **Because you multiply each height by the same base of ½, you can save some time by doing the computation like this:**

$$\frac{1}{2}\,(\ln 1.5 + \ln 2 + \ln 2.5 + \ln 3 + \ln 3.5 +$$
$$\ln 4 + \ln 4.5 + \ln 5 + \ln 5.5 + \ln 6)$$

$$\approx \frac{1}{2}\,(.405 + .693 + .916 + 1.099 + 1.253 +$$
$$1.386 + 1.504 + 1.609 + 1.705 + 1.792)$$

$$\approx \frac{1}{2}(12.362)$$

$$\approx 6.181$$

1. **a.** Estimate the area under $f(x) = \ln x$ from 1 to 6 (as in the example), but this time with 10 *left* rectangles.

   b. How is this approximation related to the area obtained with 10 right rectangles? *Hint:* Compare individual rectangles from both estimates.

   Solve It

2. Approximate the same area again with 10 midpoint rectangles.

   Solve It

3. Rank the approximations from the example and problems 1 and 2 from best to worst and defend your ranking. Obviously, you're not allowed to cheat by first finding the exact area with your calculator.

   Solve It

4. Use 8 left, right, and midpoint rectangles to approximate the area under $\sin x$ from 0 to π.

   Solve It

Sigma Notation and Reimann Sums: Geek Stuff

Now that you're warmed up, let's segue into summing some sophisticated sigma sums. *Sigma notation* may look difficult, but it's really just a shorthand way of writing a long sum.

In a sigma sum problem, you can pull anything through the sigma symbol to the outside except for a function of the *index of summation* (the *i* in the following example). Note that you can use any letter you like for the index of summation, though *i* and *k* are customary.

Q. Evaluate $\sum_{i=4}^{12} 5i^2$.

A. **The sum is 3180.**

1. **Pull the 5 through the sigma symbol.**

$5\sum_{i=4}^{12} i^2$

2. **Plug 4 into *i*, then 5, then 6, and so on up to 12, adding up all the terms.**

$= 5\left(4^2 + 5^2 + 6^2 + 7^2 + 8^2 + 9^2 + 10^2 + 11^2 + 12^2\right)$

3. **Finish on your calculator.**

$= 5(636) = 3180$

Q. Express $50^3 + 60^3 + 70^3 + 80^3 + \ldots + 150^3$ with sigma notation.

A. $1000\sum_{i=1}^{11}(i+4)^3$

1. **Create the argument of the sigma function.**

The jump amount between terms in a long sum will become the coefficient of the index of summation in a sigma sum, so you know that 10*i* is the basic term of your argument. You want to cube each term, so that gives you the following argument.

$\sum(10i)^3$

2. **Set the range of the sum.**

Ask yourself what *i* must be to make the first term equal 50^3 — that's 5, of course. And ask the same question about the last term of 150^3 — that gives you $i = 15$:

$\sum_{i=5}^{15}(10i)^3$

3. **Simplify.**

$= \sum_{i=5}^{15} 10^3 i^3$

$= 1000\sum_{i=5}^{15} i^3$

4. **(Optional) Set the *i* to begin at zero or one.**

It's often desirable to have *i* begin at 0 or 1. To turn the 5 into a 1, you subtract 4. Then subtract 4 from the 15 as well. To compensate for this subtraction, you *add* 4 to the *i* in the argument:

$= 1000\sum_{i=1}^{11}(i+4)^3$

Q. Estimate the area under $f(x) = x^2 + 3x$ from 0 to 5 using 20 right rectangles. Use sigma notation where appropriate. Then use sigma notation to express the area approximation when you use n rectangles.

A. For 20 rectangles: ≈ 84.2; for n rectangles:

$$\frac{475n^2 + 600n + 125}{6n^2}$$

1. Sketch the function and the first few and the last right rectangles. See the following figure.

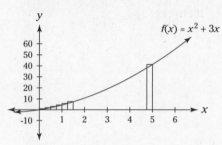

$f(x) = x^2 + 3x$

2. Add up the area of 20 rectangles. Each has an area of *base* times *height*. So for starters you've got

$$\sum_{20 \text{ rectangles}} (base \cdot height)$$

3. Plug in the base and height information to get your sigma summation.

The base of each rectangle is $\frac{(5-0)}{20}$, or $\frac{1}{4}$.

So you've got $\sum_{20} \frac{1}{4} height = \frac{1}{4} \sum_{20} height$.

The height of the first rectangle is $f\left(\frac{1}{4}\right)$, the second is $f\left(\frac{2}{4}\right)$, the third is $f\left(\frac{3}{4}\right)$, and so on until the last rectangle, which has a height of $f(5)$. This is where the index, i, comes in. You can see that the jump amount from term to term is $\frac{1}{4}$, so the argument will contain a $\frac{1}{4}i$:

$$\frac{1}{4}\sum_{20} f\left(\frac{1}{4}i\right).$$

4. Create the sum range.

i has to equal 1 to make the first term $f\left(\frac{1}{4}\right)$. And because you've got to add up 20 rectangles, i has to run through 20 numbers, so it goes from 1 to 20:

$$\frac{1}{4}\sum_{i=1}^{20} f\left(\frac{1}{4}i\right).$$

5. Replace the general function expression with your specific function, $f(x) = x^2 + 3x$.

$$\frac{1}{4}\sum_{i=1}^{20}\left[\left(\frac{1}{4}i\right)^2 + 3\left(\frac{1}{4}i\right)\right]$$

6. Simplify, pulling everything to the outside, except functions of i.

$$= \frac{1}{4}\sum_{i=1}^{20}\left(\frac{1}{4}i\right)^2 + \frac{1}{4}\sum_{i=1}^{20} 3\left(\frac{1}{4}i\right)$$

$$= \frac{1}{4}\sum_{i=1}^{20}\frac{1}{16}i^2 + \frac{1}{4}\sum_{i=1}^{20}\frac{3}{4}i$$

$$= \frac{1}{64}\sum_{i=1}^{20}i^2 + \frac{3}{16}\sum_{i=1}^{20}i$$

7. Compute the area, using the following rules for summing consecutive integers and consecutive squares of integers.

The sum of the first n integers equals $\frac{n(n+1)}{2}$, and the sum of the squares of the first n integers equals $\frac{n(n+1)(2n+1)}{6}$.

So now you've got:

$$\frac{1}{64}\left(\frac{20(20+1)(2\cdot20+1)}{6}\right) + \frac{3}{16}\left(\frac{20(20+1)}{2}\right)$$

$$= \frac{1}{64}\left(\frac{20\cdot21\cdot41}{6}\right) + \frac{3}{16}\cdot10\cdot21$$

$$= \frac{17220}{384} + \frac{630}{16}$$

$$\approx 84.2$$

8. Express the sum of n rectangles instead of 20 rectangles.

Look back at Step 5. The $\frac{1}{4}$ outside and the two $\frac{1}{4}$s inside come from the width of the rectangles that you got by dividing 5 (the span) by 20. So the width of each rectangle could have been written as $\frac{5}{20}$. To add n rectangles instead of 20, just replace the 20 with an n — that's $\frac{5}{n}$. So the three $\frac{1}{4}$s become $\frac{5}{n}$. At the same time, replace the 20 on top of the $\sum$ with an n:

$$\frac{5}{n}\sum_{i=1}^{n}\left[\left(\frac{5}{n}i\right)^2 + 3\left(\frac{5}{n}i\right)\right]$$

continued

9. **Simplify as in Step 6.**

$$= \frac{5}{n}\sum_{i=1}^{n}\left(\frac{5}{n}i\right)^{2} + \frac{5}{n}\sum_{i=1}^{n}3\left(\frac{5}{n}i\right)$$

$$= \frac{5}{n}\sum_{i=1}^{n}\frac{25}{n^{2}}i^{2} + \frac{5}{n}\sum_{i=1}^{n}\frac{15}{n}i$$

$$= \frac{125}{n^{3}}\sum_{i=1}^{n}i^{2} + \frac{75}{n^{2}}\sum_{i=1}^{n}i$$

10. **Now replace the sigma sums with the expressions for the sums of integers and squares of integers like you did in Step 7.**

$$= \frac{125}{n^{3}}\left(\frac{n(n+1)(2n+1)}{6}\right) + \frac{75}{n^{2}}\left(\frac{n(n+1)}{2}\right)$$

$$= \frac{250n^{2}+375n+125}{6n^{2}} + \frac{75n^{2}+75n}{2n^{2}}$$

$$= \frac{475n^{2}+600n+125}{6n^{2}}$$

Done! Finally! That's the formula for approximating the area under $f(x) = x^{2} + 3x$ from 0 to 5 with n rectangles — the more you use, the better your estimate. I bet you can't wait to do one of these problems on your own.

Check this result by plugging 20 into n to see whether you get the same answer as with the 20-rectangle version of this problem.

$$= \frac{475(20)^{2}+600(20)+125}{6(20)^{2}} \approx 84.2$$

It checks.

---

5. Evaluate $\sum_{i=1}^{10}4$.

Solve It

6. Evaluate $\sum_{i=0}^{9}(-1)^{i}(i+1)^{2}$.

Solve It

7. Evaluate $\sum_{i=1}^{50}(3i^2 + 2i)$.

Solve It

8. Express the following sum with sigma notation: $30 + 35 + 40 + 45 + 50 + 55 + 60$.

Solve It

9. Express the following sum with sigma notation: $8 + 27 + 64 + 125 + 216$.

Solve It

10. Use sigma notation to express the following:
$-2 + 4 - 8 + 16 - 32 + 64 - 128 + 256 - 512 + 1024$

Solve It

*11. Use sigma notation to express an 8-right-rectangle approximation of the area under $g(x) = 2x^2 + 5$ from 0 to 4. Then compute the approximation.

Solve It

*12. Using your result from problem 11, write a formula for approximating the area under g from 0 to 5 with n rectangles.

Solve It

Close Isn't Good Enough: The Definite Integral and Exact Area

Now, finally, the first calculus in this chapter. Why settle for approximate areas when you can use the definite integral to get exact areas?

The exact area under a curve between a and b is given by the *definite integral*, which is defined as follows:

$$\int_a^b f(x)\,dx = \lim_{n \to \infty} \sum_{i=1}^{n} \left[f(x_i) \cdot \left(\frac{b-a}{n} \right) \right]$$

In plain English, this simply means that you can calculate the exact area under a curve between two points by using the kind of formula you got in Step 10 of the previous example and then taking the limit of that formula as n approaches infinity. (Okay, so maybe that wasn't *plain*, but at least it was English.)

The function inside the definite integral is called the *integrand*.

Q. The answer for the example in the last section gives the approximate area under $f(x) = x^2 + 3x$ from 0 to 5 given by n rectangles as $\dfrac{475n^2 + 600n + 125}{6n^2}$. For 20 rectangles, you found the approximate area of ~84.2. With this formula and your calculator, compute the approximate area given by 50, 100, 1000, and 10,000 rectangles, then use the definition of the definite integral to compute the exact area.

A. **The exact area is 79.1$\overline{6}$.**

$$Area_{50R} = \frac{475 \cdot 50^2 + 600 \cdot 50 + 125}{6 \cdot 50^2}$$

$$= 81.175$$

$$Area_{100R} \approx 80.169$$

$$Area_{1000R} \approx 79.267$$

$$Area_{10000R} \approx 79.177$$

These estimates are getting better and better; they appear to be headed toward something near 79. Now for the magic of calculus — actually (sort of) adding up an infinite number of rectangles.

$$\int_a^b f(x)\,dx = \lim_{n \to \infty} \sum_{i=1}^n \left[f(x_i) \cdot \left(\frac{b-a}{n} \right) \right]$$

$$\int_0^5 (x^2 + 3x)\,dx = \lim_{n \to \infty} \left(\frac{475n^2 + 600n + 125}{6n^2} \right)$$

$$= \frac{475}{6}$$

$$= 79.1\overline{6} \ \text{ or } \ 79\frac{1}{6}$$

The answer of $\dfrac{475}{6}$ follows immediately from the horizontal asymptote rule (see Chapter 4). You can also break the fraction in line two above into three pieces and do the limit the long way:

$$= \lim_{x \to \infty} \left(\frac{475n^2}{6n^2} + \frac{600n}{6n^2} + \frac{125}{6n^2} \right)$$

$$= \lim_{x \to \infty} \frac{475}{6} + \lim_{x \to \infty} \frac{100}{n} + \lim_{x \to \infty} \frac{125}{6n^2}$$

$$= \frac{475}{6} + 0 + 0$$

$$= \frac{475}{6}$$

---

13. In problem 11, you estimate the area under $g(x) = 2x^5 + 5$ from 0 to 4 with 8 rectangles. The result is 71 square units.

  a. Use your result from problem 12 to approximate the area under g with 50, 100, 1000, and 10,000 rectangles.

  b. Now use your result from problem 12 and the definition of the definite integral to determine the *exact* area under $2x^2 + 5$ from 0 to 4.

Solve It

14. a. Given the following formulas for n left, right, and midpoint rectangles for the area under $x^2 + 1$ from 0 to 3, approximate the area with 50, 100, 1000, and 10,000 rectangles with each of the three formulas:

$$L_{nR} = \frac{24n^2 - 27n + 9}{2n^2}$$

$$R_{nR} = \frac{24n^2 + 27n + 9}{2n^2}$$

$$M_{nR} = \frac{48n^2 - 9}{4n^2}$$

  b. Use the definition of the definite integral with each of three formulas from the first part of the problem to determine the exact area under $x^2 + 1$ from 0 to 3.

Solve It

Finding Area with the Trapezoid Rule and Simpson's Rule

To close this chapter, I give you two more ways to approximate an area. You can use these methods when finding the exact area is impossible. (Just take my word for it that there are functions that can't be handled with ordinary integration.) With the trapezoid rule, you draw trapezoids under the curve instead of rectangles. See Figure 9-1, which is the same function I used for the first example in this chapter.

Figure 9-1:
Ten trapezoids (actually, one's a triangle, but it works exactly like a trapezoid).

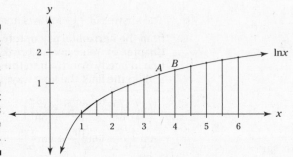

Note: You can't actually see the trapezoids, because their tops mesh with the curve, $y = \ln x$. But between each pair of points, such as A and B, there's a straight trapezoid top in addition to the curved piece of $y = \ln x$.

The Trapezoid Rule: You can approximate the exact area under a curve between a and b, $\int_a^b f(x)\, dx$, with a sum of trapezoids given by the following formula. In general, the more trapezoids, the better the estimate.

$$T_n = \frac{b-a}{2n}\Big[f(x_0) + 2f(x_1) + 2f(x_2) + 2f(x_3) + \ldots + 2f(x_{n-1}) + f(x_n)\Big]$$

where n is the number of trapezoids, x_0 equals a, and x_1 through x_n are the equally-spaced x-coordinates of the right edges of trapezoids 1 through n.

Simpson's Rule also uses trapezoid-like shapes, except that the top of each "trapezoid" — instead of being a straight-slanting segment, as "shown" in Figure 9-1 — is a curve (actually a small piece of a parabola) that very closely hugs the function. Because these little parabola pieces are so close to the function, Simpson's rule gives the best area approximation of any of the methods. If you're wondering why you should learn the trapezoid rule when you can just as easily use Simpson's rule and get a more accurate estimate, chalk it up to just one more instance of the sadism of calculus teachers.

Simpson's Rule: You can approximate the exact area under a curve between a and b, $\int_a^b f(x)\, dx$, with a sum of parabola-topped "trapezoids," given by the following formula. In general, the more "trapezoids," the better the estimate.

$$S_n = \frac{b-a}{3n}\Big[f(x_0) + 4f(x_1) + 2f(x_2) + 4f(x_3) + 2f(x_4) + \ldots + 4f(x_{n-1}) + f(x_n)\Big]$$

where n is *twice* the number of "trapezoids" and x_0 through x_n are the $n + 1$ evenly spaced x-coordinates from a to b.

Q. Estimate the area under $f(x) = \ln x$ from 1 to 6 with 10 trapezoids. Then compute the percent error.

A. **The approximate area is 5.733. The error is about 0.31%.**

1. **Sketch the function and the 10 trapezoids.**

   Already done — Figure 9-1.

2. **List the values for *a*, *b*, and *n*, and determine the 11 *x*-values, x_0 through x_{10} (the left edge of the first trapezoid plus the 10 right edges of the 10 trapezoids).**

   Note that in this and all similar problems, *a* equals x_0 and *b* equals x_n (x_{10} here).

   $a = 1$

   $b = 6$

   $n = 10$

   $x_0 = 1, \; x_1 = 1.5, \; x_2 = 2, \; x_3 = 2.5, \; \ldots \; x_{10} = 6$

3. **Plug these values into the trapezoid rule formula and solve.**

   $$T_{10} = \frac{6-1}{2 \cdot 10}(\ln 1 + 2\ln 1.5 + 2\ln 2 + 2\ln 2.5 +$$
   $$2\ln 3 + 2\ln 3.5 + 2\ln 4 + 2\ln 4.5 +$$
   $$2\ln 5 + 2\ln 5.5 + \ln 6)$$

   $$\approx \frac{5}{20}(0 + .811 + 1.386 + 1.833 + 2.197 +$$
   $$2.506 + 2.773 + 3.008 + 3.219 +$$
   $$3.409 + 1.792)$$

   $$\approx 5.733$$

4. **Compute the percent error.**

   My TI-89 tells me that the exact area is 5.7505568153635 . . . For this problem, round that off to 5.751. The percent error is given by the error divided by the exact area. So that gives you:

   $$percent\ error \approx \frac{5.751 - 5.733}{5.751} \approx .0031 = .31\%$$

   Compare this to the 10-midpoint-rectangle error we compute in the solution to problem 2: 0.14%. In general, the error with a trapezoid estimate is roughly twice the corresponding midpoint-rectangle error.

Q. Estimate the area under $f(x) = \ln x$ from 1 to 6 with 10 Simpson rule "trapezoids." Then compute the percent error.

A. **The approximate area is 5.751. The error is a mere 0.00069%.**

1. **List the values for *a*, *b*, and *n*, and determine the 21 *x*-values x_0 through x_{20}, the 11 edges and the 10 base midpoints of the 10 curvy-topped "trapezoids."**

   $a = 1$

   $b = 6$

   $n = 20$

   $x_0 = 1, \; x_1 = 1.25, \; x_2 = 1.5, \; x_3 = 1.75, \; \ldots \; x_{20} = 6$

2. **Plug these values into the formula.**

   $$S_{20} = \frac{6-1}{3 \cdot 20}(\ln 1 + 4\ln 1.25 + 2\ln 1.5 + 4\ln 1.75 +$$
   $$2\ln 2 + \ldots + 4\ln 5.75 + \ln 6)$$

   $$\approx \frac{5}{60}(69.006202893232)$$

   $$\approx 5.7505169$$

3. **Figure the percent error.**

   The exact answer, again, is 5.7505568153635. Round that off to 5.7505568.

   $$percent\ error \approx \frac{5.7505568 - 5.7505169}{5.7505568} \approx$$
   $$.0000069 = .00069\%$$

   — way better than either the midpoint or trapezoid estimate. Impressed?

15. Continuing with problem 4, estimate the area under $y = \sin x$ from 0 to π with 8 trapezoids, and compute the percent error.

Solve It

16. Estimate the same area as problem 15 with 16 and 24 trapezoids and compute the percent errors.

Solve It

17. Approximate the same area as problem 15 with 8 Simpson's rule "trapezoids" and compute the percent error.

Solve It

18. Use the following shortcut to figure S_{20} for the area under $\ln x$ from 1 to 6. (Use the results from problem 2 and the first example in this section.)

Shortcut: If you know the midpoint and trapezoid estimates for n rectangles, you can easily compute the Simpson's rule estimates for n curvy-topped "trapezoids" with the following formula:

$$S_{2n} = \frac{M_n + M_n + T_n}{3}$$

Solve It

Solutions to Getting into Integration

1   **a.** Estimate the area under $f(x) = \ln x$ from 1 to 6, but this time with 10 *left* rectangles. **The area is 5.285.**

   1. Sketch a graph and divide the intervals into 10 subintervals.

   2. a. Draw the first left rectangle by putting your pen at the *left* end of the first base (that's at $x = 1$) and going straight up till you hit the function.

       Whoops. You're already *on* the function at $x = 1$, right? So, guess what? There is no first rectangle — or you could say it's a rectangle with a height of zero and an area of zero.

   b. Draw the "second" rectangle by putting your pen at $x = 1.5$, going straight up till you hit $f(x) = \ln x$, then go *right* till you're directly above $x = 2$, then down to the x-axis. See the following figure.

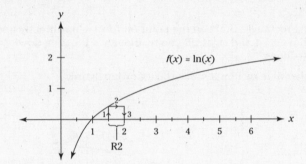

   3. Draw the rest of the rectangles. See the following figure.

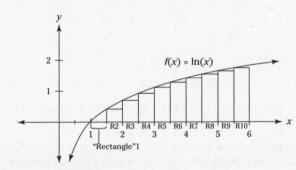

   4. Compute your approximation.

$$Area_{10\,LRs} = \frac{1}{2}(\ln 1 + \ln 1.5 + \ln 2 + \ln 2.5 + \ln 3 + \ln 3.5 + \ln 4 + \ln 4.5 + \ln 5 + \ln 5.5)$$

$$= \frac{1}{2}(0 + .405 + .693 + .916 + 1.099 + 1.253 + 1.386 + 1.504 + 1.609 + 1.705)$$

$$= \frac{1}{2}(10.57)$$

$$= 5.285$$

   b. How is this approximation related to the area obtained with 10 right rectangles? Look at the second line in the computation in Step 4. Note that the sum of the 10 numbers inside the parentheses includes the first 9 numbers in the same line in the computation for right rectangles (see the example). The only difference is that the sum for left rectangles has a 0 at the left end and the sum for right rectangles has a 1.792 at the right end.

If you look at the figure in Step 2 of the example and at the figure in Step 3 of the solution to 1(a), you'll see why this works out this way. The first rectangle in the example figure is identical to the second rectangle in the solution 1(a) figure. The second rectangle in the example figure is identical to the third rectangle in the solution 1(a) figure, and so on. The only difference is that the solution 1(a) figure contains the left-most "rectangle" (the invisible one) and the example figure contains the tall, right-most rectangle.

A left-rectangle sum and a right-rectangle sum will always differ by an amount equal to the difference in area of the left-most left rectangle and the right-most right rectangle. (Memorize this sentence and recite it in class — with your right index finger pointed upward for effect. You'll instantly become a babe (dude) magnet.)

2 Approximate the same area again with 10 midpoint rectangles. **The approximate area is 5.759.**

1. Sketch your curve and the 10 subintervals again.

2. Compute the midpoints of the bases of all rectangles. This should be a no-brainer: 1.25, 1.75, 2.25 . . . 5.75.

3. Draw the first rectangle. Start on the point on $f(x) = \ln x$ directly above $x = 1.25$, then go left till you're above $x = 1$ and right till you're above $x = 1.5$, then down from both these points to make the two sides.

4. Draw the other nine rectangles. See the following figure.

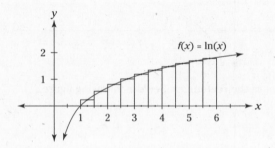

5. Compute your estimate.

$$Area_{10 \text{ MRs}} = \frac{1}{2}(\ln 1.25 + \ln 1.75 + \ln 2.25 + \ln 2.75 + \ln 3.25 + \ln 3.75 + \ln 4.25 + \ln 4.75 + \ln 5.25 + \ln 5.75)$$

$$\approx \frac{1}{2}(.223 + .560 + .811 + 1.012 + 1.179 + 1.322 + 1.447 + 1.558 + 1.658 + 1.749)$$

$$\approx 5.760$$

3 Rank the approximations from the example and problems 1 and 2 from best to worst and defend your ranking. **The midpoint rectangles give the best estimate because each rectangle goes above the curve (in this sense, it's too big) and also leaves an uncounted gap below the curve (in this sense, it's too small). These two errors cancel each other out to some extent. By the way, the exact area is about 5.751. The approximate area with 10 midpoint rectangles of 5.759 is only about 0.14% off.**

It's harder to rank the left versus the right rectangle estimates. Kudos if you noticed that because of the shape of $f(x) = \ln x$ (technically because it's concave *down* and *increasing*), right rectangles will give a slightly better estimate. It turns out that the right-rectangle approximation is off by 7.48% and the left-rectangle estimate is off by 8.10%. If you missed this question, don't sweat it. It's basically an extra-credit type question.

4 Use 8 left, right, and midpoint rectangles to approximate the area under $\sin x$ from 0 to π.

Let's cut to the chase. Here are the computations for 8 left rectangles, 8 right rectangles, and 8 midpoint rectangles:

$$Area_{8\,LR} = \frac{\pi}{8}\left(\sin 0 + \sin \frac{\pi}{8} + \sin \frac{2\pi}{8} + \sin \frac{3\pi}{8} + \sin \frac{4\pi}{8} + \sin \frac{5\pi}{8} + \sin \frac{6\pi}{8} + \sin \frac{7\pi}{8}\right)$$

$$= \frac{\pi}{8}(0 + .383 + .707 + .924 + 1 + .924 + .707 + .383) = \frac{\pi}{8}(5.027) = \mathbf{1.974}$$

$$Area_{8\,RR} = \frac{\pi}{8}\left(\sin \frac{\pi}{8} + \sin \frac{2\pi}{8} + \sin \frac{3\pi}{8} + \sin \frac{4\pi}{8} + \sin \frac{5\pi}{8} + \sin \frac{6\pi}{8} + \sin \frac{7\pi}{8} + \sin \pi\right)$$

$$= \frac{\pi}{8}(.383 + .707 + .924 + 1 + .924 + .707 + .383 + 0) = \frac{\pi}{8}(5.027) = \mathbf{1.974}$$

$$Area_{8\,MR} = \frac{\pi}{8}\left(\sin \frac{\pi}{16} + \sin \frac{3\pi}{16} + \sin \frac{5\pi}{16} + \sin \frac{7\pi}{16} + \sin \frac{9\pi}{16} + \sin \frac{11\pi}{16} + \sin \frac{13\pi}{16} + \sin \frac{15\pi}{16}\right)$$

$$= \frac{\pi}{8}(.195 + .556 + .831 + .981 + .981 + .831 + .556 + .195) = \frac{\pi}{8}(5.126) = \mathbf{2.013}$$

The exact area under $\sin x$ from 0 to π has the wonderfully simple answer of 2. The error of the midpoint rectangle estimate is 0.65%, and the other two have an error of 1.3%. The left and right rectangle estimates are the same, by the way, because of the symmetry of the sine wave.

5 $\sum_{i=1}^{10} 4 = \mathbf{40}$

As often happens with many types of problems in mathematics, this very simple version of a sigma sum problem is surprisingly tricky. Here, there's no place to plug in the i, so all the i does is work as a counter:

$$\sum_{i=1}^{10} 4 = 4 + 4 + 4 + 4 + 4 + 4 + 4 + 4 + 4 + 4 = 10 \cdot 4 = 40$$

6 $\sum_{i=0}^{9} (-1)^i (i+1)^2 = \mathbf{-55}$

$$= (-1)^0 (0+1)^2 + (-1)^1 (1+1)^2 + (-1)^2 (2+1)^2 + \ldots$$
$$= 1^2 - 2^2 + 3^2 - 4^2 + 5^2 - 6^2 + 7^2 - 8^2 + 9^2 - 10^2$$
$$= -55$$

7 $\sum_{i=1}^{50} (3i^2 + 2i) = \mathbf{131,325}$

$$= \sum_{i=1}^{50} 3i^2 + \sum_{i=1}^{50} 2i = 3\sum_{i=1}^{50} i^2 + 2\sum_{i=1}^{50} i$$
$$= 3\left(\frac{50(50+1)(2 \cdot 50 + 1)}{6}\right) + 2\left(\frac{50(50+1)}{2}\right) = 131,325$$

8 $30 + 35 + 40 + 45 + 50 + 55 + 60 = \sum_{k=6}^{12} \mathbf{5k}$ **or** $\sum_{k=1}^{7} \mathbf{5}(\mathbf{k}+\mathbf{5})$ **or** $\sum_{k=1}^{7} (\mathbf{5k}+\mathbf{25})$

9 $8 + 27 + 64 + 125 + 216 = \sum_{k=2}^{6} \mathbf{k^3}$ **or** $\sum_{k=1}^{5} (\mathbf{k}+\mathbf{1})^3$ Did you recognize this pattern?

10 $-2 + 4 - 8 + 16 - 32 + 64 - 128 + 256 - 512 + 1024 = \sum_{i=1}^{10} (-1)^i 2^i$ **or** $\sum_{i=1}^{10} (-2)^i$

To make the terms in a sigma sum alternate between positive and negative, use a −1 raised to a power in the argument. The power will usually be i or $i + 1$.

*11 Use sigma notation to express an 8-right-rectangle approximation of the area under $g(x) = 2x^2 + 5$ from 0 to 4. Then compute the approximation. **The notation and approximation are** $\frac{1}{4}\sum\limits_{i=1}^{8} i^2 + 20 = 71$.

1. Sketch $g(x)$. You're on your own.

2. Express the basic idea of your sum: $\sum\limits_{8 \text{ rectangles}} (base \cdot height)$.

3. Figure the base and plug in.

$$base = \frac{4-0}{8} = \frac{1}{2}$$

$$\sum\limits_{8}\left(\frac{1}{2} \cdot height\right) = \frac{1}{2}\sum\limits_{8} height$$

4. Express the height as a function of the index of summation, and add the limits of summation:

$$\frac{1}{2}\sum\limits_{i=1}^{8} f\left(\frac{1}{2}i\right)$$

5. Plug in your function, $g(x) = 2x^2 + 5$.

$$= \frac{1}{2}\sum\limits_{i=1}^{8}\left[2\left(\frac{1}{2}i\right)^2 + 5\right]$$

6. Simplify: $= \frac{1}{2}\sum\limits_{i=1}^{8} 2\left(\frac{1}{2}i\right)^2 + \frac{1}{2}\sum\limits_{i=1}^{8} 5 = \sum\limits_{i=1}^{8}\left(\frac{1}{2}\right)^2 i^2 + \frac{1}{2} \cdot 40 = \frac{1}{4}\sum\limits_{i=1}^{8} i^2 + 20$

7. Use the sum of squares rule to finish: $= \frac{1}{4}\left(\frac{8(8+1)(2\cdot 8+1)}{6}\right) + 20 = 51 + 20 = 71$

*12 Using your result from problem 11, write a formula for approximating the area under g from 0 to 5 with n rectangles. **The formula is** $\dfrac{188n^2 + 192n + 64}{3n^2}$.

1. Convert the sigma formula for summing 8 rectangles to one for summing n rectangles.

Look at Step 5 from the previous solution. The number $\frac{1}{2}$ appears twice. You got $\frac{1}{2}$ when you computed the width of the base of each rectangle. That's $\frac{4-0}{8}$, or $\frac{4}{8}$. You want a formula for n rectangles instead of 8, so use $\frac{4}{n}$ instead of $\frac{1}{2}$ and replace the 8 on top of $\sum$ with an n.

$$\frac{4}{n}\sum\limits_{i=1}^{n}\left[2\left(\frac{4}{n}i\right)^2 + 5\right]$$

2. Simplify: $= \frac{4}{n}\sum\limits_{i=1}^{n} 2\left(\frac{4}{n}i\right)^2 + \frac{4}{n}\sum\limits_{i=1}^{n} 5 = \frac{4}{n}\sum\limits_{i=1}^{n}\left(2\cdot\frac{16}{n^2}\cdot i^2\right) + \frac{4}{n}\cdot 5n = \frac{128}{n^3}\sum\limits_{i=1}^{n} i^2 + 20$

3. Use the sum of squares formula.

$$= \frac{128}{n^3}\left(\frac{n(n+1)(2n+1)}{6}\right) + 20 = \frac{128n^2 + 192n + 64}{3n^2} + 20 = \frac{188n^2 + 192n + 64}{3n^2}$$

13 **a.** Use your result from problem 12 to approximate the area with 50, 100, 1000 and 10,000 rectangles.

$$Area_{nR} = \frac{188n^2 + 192n + 64}{3n^2}$$

$$Area_{50R} = \frac{188\cdot 50^2 + 192\cdot 50 + 64}{3\cdot 50^2}$$

$$\approx \textbf{63.955}$$

Because all right-rectangle estimates with this curve will be over-estimates, this result shows how far off the approximation of 71 square units was. The answers for the rest of the approximations are

$$Area_{100R} \approx \mathbf{63.309}$$

$$Area_{1000R} \approx \mathbf{62.731}$$

$$Area_{10,000R} \approx \mathbf{62.673}$$

b. Now use your result from problem 12 and the definition of the definite integral to determine the *exact* area under $2x^2 + 5$ from 0 to 4. **The area is 62.666 ... or 62⅔.**

$$\int_a^b f(x)\,dx = \lim_{n \to 0} \sum_{i=1}^n \left[f(x_i)\left(\frac{b-a}{n}\right)\right]$$

$$\int_0^4 (2x^2 + 5)\,dx = \lim_{n \to \infty} \frac{188n^2 + 192n + 64}{3n^2}$$

$$= \frac{188}{3}$$

$$= 62.\overline{6} \quad \text{or} \quad 62\frac{2}{3}$$

14 **a.** Given the following formulas for left, right, and midpoint rectangles for the area under $x^2 + 1$ from 0 to 3, approximate the area with 50, 100, 1000, and 10,000 rectangles with each of the three formulas.

$L_{50R} \approx \mathbf{11.732}$	$R_{50R} \approx \mathbf{12.272}$	$M_{50R} = \mathbf{11.9991}$
$L_{100R} \approx \mathbf{11.865}$	$R_{100R} \approx \mathbf{12.135}$	$M_{100R} = \mathbf{11.999775}$
$L_{1000R} \approx \mathbf{11.987}$	$R_{1000R} \approx \mathbf{12.014}$	$M_{1000R} = \mathbf{11.99999775}$
$L_{10,000R} \approx \mathbf{11.999}$	$R_{10,000R} \approx \mathbf{12.001}$	$M_{10,000R} = \mathbf{11.9999999775}$

You can see from the results how much better the midpoint-rectangle estimates are than the other two.

b. Use the definition of the definite integral with each of three formulas from the first part of the problem to determine the exact area under $x^2 + 1$ from 0 to 3.

For left rectangles, $\displaystyle\int_0^3 (x^2 + 1)\,dx = \lim_{n \to \infty} \frac{24n^2 - 27n + 9}{2n^2} = \frac{24}{2} = \mathbf{12}$

For right rectangles, $\displaystyle\int_0^3 (x^2 + 1)\,dx = \lim_{n \to \infty} \frac{24n^2 + 27n + 9}{2n^2} = \frac{24}{2} = \mathbf{12}$

And for midpoint rectangles, $\displaystyle\int_0^3 (x^2 + 1)\,dx = \lim_{n \to \infty} \frac{48n^2 - 9}{4n^2} = \frac{48}{4} = \mathbf{12}$

Big surprise — they all equal 12. They better all come out the same since you're computing the *exact* area.

15 Continuing with problem 4, estimate the area under $y = \sin x$ from 0 to π with eight trapezoids, and compute the percent error. **The approximate area is 1.974 and the error is 1.3%.**

1. List the values for a, b, and n, and determine the x-values x_0 through x_8.

$a = 0$

$b = \pi$

$n = 8$

$x_0 = 0, \ x_1 = \frac{\pi}{8}, \ x_2 = \frac{2\pi}{8}, \ x_3 = \frac{3\pi}{8}, \ \dots \ x_8 = \pi$

2. Plug these values into the formula.

$$T_8 = \frac{\pi - 0}{2 \cdot 8}\left(\sin 0 + 2\sin\frac{\pi}{8} + 2\sin\frac{2\pi}{8} + 2\sin\frac{3\pi}{8} + \ldots + 2\sin\frac{7\pi}{8} + \sin\pi\right)$$

$$\approx \frac{\pi}{16}(0 + .765 + 1.414 + 1.848\ldots + 0) \approx 1.974$$

The exact area of 2 was given in problem 4, and thus the percent error is $(2 - 1.974)/2$, or 1.3%.

16 Estimate the same area with 16 and 24 trapezoids and compute the percent error.

$$T_{16} = \frac{\pi - 0}{2 \cdot 16}\left(\sin 0 + 2\sin\frac{\pi}{16} + 2\sin\frac{2\pi}{16} + 2\sin\frac{3\pi}{16} + \ldots + 2\sin\frac{15\pi}{16} + \sin\pi\right)$$

$$\approx \frac{\pi}{32}(0 + .390 + .765 + \ldots + 0) \approx \mathbf{1.994}$$

The percent error for 16 trapezoids is about 0.3%.

$$T_{24} = \frac{\pi - 0}{2 \cdot 24}\left(\sin 0 + 2\sin\frac{\pi}{24} + 2\sin\frac{2\pi}{24} + 2\sin\frac{3\pi}{24} + \ldots + 2\sin\frac{23\pi}{24} + \sin\pi\right)$$

$$\approx \frac{\pi}{48}(0 + .261 + .518 + \ldots + 0) \approx \mathbf{1.997}$$

The percent error for 24 trapezoids is about 0.15%.

17 Approximate the same area with eight Simpson's rule "trapezoids" and compute the percent error. **The area for 8 "trapezoids" is 2.00001659 and the error is 0.000830%.**

For 8 Simpson's "trapezoids":

1. List the values for a, b, and n, and determine the x-values x_0 through x_{16}, the 9 edges and the 8 base midpoints of the 8 curvy-topped "trapezoids."

   $a = 0$

   $b = \pi$

   $n = 16$

   $x_0 = 0,\ x_1 = \frac{\pi}{16},\ x_2 = \frac{2\pi}{16},\ x_3 = \frac{3\pi}{16},\ \ldots\ x_{16} = \pi$

2. Plug these values into the formula.

$$S_{16} = \frac{\pi - 0}{3 \cdot 16}\left(\sin 0 + 4\sin\frac{\pi}{16} + 2\sin\frac{2\pi}{16} + 4\sin\frac{3\pi}{16} + 2\sin\frac{4\pi}{16} + \ldots + 4\sin\frac{15\pi}{16} + \sin\pi\right)$$

$$\approx \frac{\pi}{48}(0 + .7804 + .7654 + 2.2223 + 1.4142 + \ldots + 0.7804 + 0) \approx 2.00001659$$

The percent error for eight Simpson "trapezoids" is 0.000830%.

18 Use the following shortcut to figure S_{20} for the area under $\ln x$ from 1 to 6.

Using the formula in the problem, you get:

$$S_{2n} = \frac{M_n + M_n + T_n}{3}$$

$$S_{20} = \frac{M_{10} + M_{10} + T_{10}}{3}$$

$$\approx \frac{5.759 + 5.759 + 5.733}{3}$$

$$\approx 5.750$$

This agrees (except for a small round-off error) with the result obtained the hard way in the Simpson's rule example problem.

Chapter 10

Integration: Reverse Differentiation

. . . .

In This Chapter

▶ Analyzing the area function

▶ Getting off your fundament (butt) to study the Fundamental Theorem

▶ Guessing and checking

▶ Pulling the switcheroo

. . . .

1n this chapter, you really get into integration in full swing. First you look at the annoying area function, then the Fundamental Theorem of Calculus, and then two beginner integration methods.

The Absolutely Atrocious and Annoying Area Function

The area function is both more difficult and less useful than the material that follows it. With any luck, your calc teacher will skip it or just give you a cursory introduction to it. Once you get to the following section on the Fundamental Theorem of Calculus, you'll have no more use for the area function. It's taught because it's the foundation for the all-important Fundamental Theorem.

The area function is an odd duck and doesn't look like any function you've ever seen before.

$$A_f(x) = \int_s^x f(t)\, dt$$

The input of the function (its *argument*) is the x on top of the integral symbol. Note that $f(t)$ is *not* the argument. The output, $A_f(x)$, tells you how much area has been swept out under the curve, $f(t)$, between some starting point, $x = s$, and the input value. For example, consider the simple horizontal line $g(t) = 10$ and the area function based on it, $A_g(x) = \int_3^x 10\, dt$.

This area function tells you how much area is under the horizontal line between 3 and the input value. When $x = 4$, the area is 10 because you've got a rectangle with a base of one — from 3 to 4 — and a height of 10. When $x = 5$, the output of the function is 20; when $x = 6$, the output is 30, and so on. (For an excellent and thorough explanation of the area function and how it relates to the Fundamental Theorem, check out *Calculus For Dummies*.) The best way to get a handle on this weird function is to see it in action, so here goes.

Don't forget that when using an area function (or a definite integral — stay tuned), area below the x-axis counts as *negative* area.

EXAMPLE

Q. Consider $f(t)$, shown in the following figure. Given the area function

$$A_f(x) = \int_{2}^{x} f(t)\, dt,$$ approximate $A_f(4)$,

$A_f(5)$, $A_f(2)$, and $A_f(0)$. Is A_f increasing or decreasing between $x = 5$ and $x = 6$? Between $x = 8$ and $x = 9$?

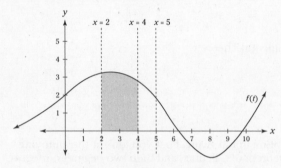

A. $A_f(4)$ **is the area under** $f(t)$ **between 2 and 4. That's roughly a rectangle with a base of 2 and a height of 3, so the area is about 6. (See the shaded area in the figure.)**

$A_f(5)$ adds a bit to $A_f(4)$ — roughly a trapezoid with "height" of 1 and "bases" 2 and 3 (along the dotted lines at $x = 4$ and $x = 5$) that thus has an area of 2.5, so $A_f(5)$ is roughly 6 plus 2.5, or 8.5.

$A_f(2)$ is the area between 2 and 2, which is zero.

$A_f(0)$ is another area roughly in the shape of a trapezoid. Its height is 2 and its bases are 2 and 3, so its area is 5. But because you go backwards from 2 to zero, $A_f(0)$ equals –5.

Between $x = 5$ and $x = 6$, A_f is *increasing*. Be careful here: $f(t)$ is decreasing between 5 and 6, but as you go from 5 to 6, A_f sweeps out more and more area so it's increasing.

Between $x = 8$ and $x = 9$, while $f(t)$ is increasing A_f is *decreasing*. Area below the x-axis counts as negative area, so in moving from 8 to 9, A_f sweeps out more and more negative area, thus growing more and more negative, and thus A_f is decreasing.

1. For problems 1 through 4, use the area function $A_g(x) = \int_{1/2}^{x} g(t)\, dt$ and the following figure. Most answers will be approximations. Where (from 0 to 8) does A_g equal 0?

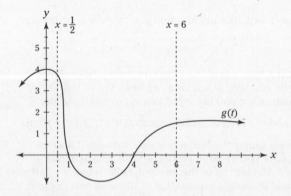

Solve It

2. Where (from $x = 0$ to $x = 8$) does A_g reach

a. its maximum value?

b. its minimum value?

Solve It

3. In what intervals between 0 and 8 is A_g

   a. increasing?

   b. decreasing?

Solve It

4. Approximate $A_g(1)$, $A_g(3)$, and $A_g(5)$.

Solve It

Sound the Trumpets: The Fundamental Theorem of Calculus

The absolutely incredibly fantastic Fundamental Theorem of Calculus — some say one of or perhaps *the* greatest theorem in the history of mathematics — gives you a neat shortcut for finding area so you don't have to deal with the annoying area function or that rectangle mumbo jumbo from Chapter 9. The basic idea is that you use the *antiderivative* of a function to find the area under it.

Let me jog your memory on antiderivatives: Because $3x^2$ is the derivative of x^3, x^3 is an *antiderivative* of $3x^2$. But so is $x^3 + 5$ because its derivative is also $3x^2$. So anything of the form $x^3 + C$ (C is a constant) is an antiderivative of $3x^2$. Technically, you say that $x^3 + C$ is the *indefinite integral* of $3x^2$ and that $x^3 + C$ is the *family* of antiderivatives of $3x^2$.

The Fundamental Theorem comes in two versions: useless and useful. You learn the useless version for basically the same reason you studied geometry proofs in high school, namely, "just because."

The Fundamental Theorem of Calculus (the difficult, mostly useless version): Given an area function A_f that sweeps out area under $f(t)$, $A_f(x) = \int_s^x f(t)\,dt$, the rate at which area is being swept out is equal to the height of the original function. So, because the rate is the derivative, the derivative of the area function equals the original function: $\frac{d}{dx}A_f(x) = f(x)$. Because $A_f(x) = \int_s^x f(t)\,dt$, you can also write the previous equation as follows: $\frac{d}{dx}\int_s^x f(t)\,dt = f(x)$.

The Fundamental Theorem of Calculus (the easy, useful version): Let F be any antiderivative of the function f; then

$$\int_a^b f(x)\,dx = F(b) - F(a)$$

Q. **a.** For the area function

$$A_f(x) = \int_{10}^x (t^2 - 5t)\,dt, \text{ what's } \frac{d}{dx}A_f(x)?$$

b. For the area function, $B_f(x) = \int_{-4}^{3x^2} \sin t\,dt$, what's $\frac{d}{dx}B_f(x)$?

A. **a.** No work needed here. The answer is simply $x^2 - 5x$

b. $6x\sin 3x^2$

The *argument* of an area function is the expression at the top of the integral symbol — *not* the integrand. Because the argument of this area function, $3x^2$, is something other than a plain old x, this is a chain rule problem. Thus,

$$\frac{d}{dx}B_f(x) = \sin 3x^2 \cdot 6x, \text{ or } 6x\sin 3x^2.$$

Q. What's the area under $2x^2 + 5$ from 0 to 4? Note this is the same question you worked on in Chapter 9 with the difficult, sigma-sum-rectangle method.

A. $\frac{188}{3}$

Using the second version of the Fundamental Theorem,

$$\int_0^4 (2x^2 + 5)\,dx = F(4) - F(0) \text{ where } F$$

is any antiderivative of $2x^2 + 5$.

By trial and error, you can find that the derivative of $\frac{2}{3}x^3 + 5x$ is an antiderivative of $2x^2 + 5$. Thus,

$$\int_0^4 (2x^2 + 5)\,dx = \frac{2}{3}x^3 + 5x \Big|_0^4$$

$$= \left(\frac{2}{3}4^3 + 5 \cdot 4\right) - \left(\frac{2}{3} \cdot 0^3 + 5 \cdot 0\right)$$

$$= \frac{188}{3}$$

The same answer with *much* less work than adding up all those rectangles!

5. **a.** If $A_f(x) = \int_0^x \sin t\, dt$, what's $\frac{d}{dx} A_f(x)$?

b. If $A_g(x) = \int_{\pi/4}^x \sin t\, dt$, what's $\frac{d}{dx} A_g(x)$?

Solve It

*6. Given that $A_f(x) = \int_{-\pi/4}^{\cos x} \sin t\, dt$, find $\frac{d}{dx} A_f(x)$.

Solve It

7. For $A_f(x)$ from problem 5a, where does $\frac{d}{dx} A_f$ equal zero?

Solve It

*8. For $A_f(x)$ from problem 6, evaluate $A_f'\left(\frac{\pi}{4}\right)$.

Solve It

9. What's the area under $y = \sin x$ from 0 to π?

Solve It

10. Evaluate $\int_0^{2\pi} \sin x \, dx$.

Solve It

11. Evaluate $\int_2^3 (x^3 - 4x^2 + 5x - 10) \, dx$.

Solve It

12. Evaluate $\int_{-1}^2 e^x \, dx$.

Solve It

Finding Antiderivatives: The Guess and Check Method

Your textbook, as well as the Cheat Sheet in *Calculus For Dummies*, lists a set of antiderivatives that you should memorize, such as $\sin x$, $\frac{1}{x}$, or $\frac{1}{1+x^2}$. (Most of them are simply the basic derivative rules you know written in reverse.) When you face a problem that's similar to one of these — like finding the antiderivative of $\sin 5x$ or $\frac{1}{8x}$ — you can use the guess and check method: just guess your answer, check it by differentiating, then if it's wrong, tweak it till it works.

Q. What's $\int \sin 3x \, dx$?

A. $-\frac{1}{3}\cos 3x$

You've memorized that $-\cos x$ is the antiderivative of $\sin x$ — because, of course, $\sin x$ is the derivative of $-\cos x$. So a good guess for this antiderivative would be $-\cos 3x$. When you check that guess by taking its derivative with the chain rule, you get $3\sin 3x$, which is what you want except for that first 3. To compensate for that, simply *divide* your guess by 3: $\frac{-\cos 3x}{3}$. That's it. If you have any doubts about this second guess, take its derivate and you'll see that it gives you the desired integrand, $\sin 3x$.

13. Determine $\int (4x-1)^3 \, dx$.

Solve It

14. What's $\int \sec^2 6x \, dx$?

Solve It

15. Determine $\int \cos\frac{x-1}{2}\,dx$.

Solve It

16. What's $\int \frac{3\,dt}{2t+5}$?

Solve It

17. Compute the definite integral,

$$\int_0^\pi \frac{5}{\pi}\sec(5t-\pi)\tan(5t-\pi)\,dt.$$

Solve It

18. Antidifferentiate $\int \frac{4.5}{1+9x^2}\,dx$.

Solve It

The Substitution Method: Pulling the Switcheroo

The group of guess-and-check problems in the last section involve integrands that differ from the standard integrand of a memorized antiderivative rule by a *numerical* amount. The next set of problems involves integrands where the extra thing they contain includes a *variable* expression. For these problems, you can still use the guess-and-check method, but the traditional way of doing such problems is with the substitution method.

Q. Antidifferentiate $\int x^2 \sin x^3 \, dx$ with the substitution method.

A. $-\frac{1}{3} \cos x^3 + C$

1. If a function in the integrand has something other than a plain old x for its argument, set u equal to that argument.

$$u = x^3$$

2. Take the derivative of u with respect to x, then throw the dx to the right side.

$$\frac{du}{dx} = 3x^2$$
$$du = 3x^2 \, dx$$

3. Tweak your integrand so it contains the result from Step 2 ($3x^2 \, dx$); and compensate for this tweak amount by multiplying the integral by the reciprocal of the tweak number.

$$\int x^2 \sin x^3 \, dx$$

You need a 3 in the integrand, so put in a 3 and compensate with a $\frac{1}{3}$.

$$= \frac{1}{3} \int 3x^2 \sin x^3 \, dx$$
$$= \frac{1}{3} \int \sin \underset{u}{\underline{x^3}} \, \underset{du}{\underline{3x^2 \, dx}}$$

4. Pull the switcheroo.

$$= \frac{1}{3} \int \sin u \, du$$

5. Antidifferentiate by using the derivative of $-\cos x$ in reverse.

$$= -\frac{1}{3} \cos u + C$$

6. Get rid of the u by switching back to the original expression.

$$= -\frac{1}{3} \cos x^3 + C$$

Q. Evaluate $\int_0^{\sqrt[3]{\pi}} x^2 \sin x^3 \, dx$.

A. $\frac{2}{3}$

1. This is the same as the previous Step 1 except that at the same time as setting u equal to x^3, you take the two x-indices of integration and turn them into u-indices of integration.

Like this:

$$u = x^3$$

when $x = 0$, $u = 0$

when $x = \sqrt[3]{\pi}$, $u = \sqrt[3]{\pi}^3 = \pi$

So 0 and π are the two u-indices of integration.

2.–3. Steps 2-3 are identical to 2-3 in the previous example except that you happen to be dealing with a definite integral in this problem.

4. Pull the switcheroo. This time in addition to replacing the x^3 and the $3x^2 \, dx$ with their u-equivalents, you also replace the x-indices with the u-indices:

$$= \frac{1}{3} \int_0^{\pi} \sin u \, du$$

5. Evaluate.

$$= -\frac{1}{3} \cos u \Big]_0^{\pi}$$
$$= -\frac{1}{3}(-1 - 1) = \frac{2}{3}$$

If you prefer, you can skip determining the u-indices of integration; just replace the u with x^3 like you did above in Step 6, and then evaluate the definite integral with the original indices of integration. (Your calc teacher may not like this, however, because it's not the book method.)

$$= -\frac{1}{3} \cos x^3 \Big]_0^{\sqrt[3]{\pi}}$$
$$= -\frac{1}{3}\left(\cos \sqrt[3]{\pi}^3 - \cos 0^3\right)$$
$$= -\frac{1}{3}(-1 - 1) = \frac{2}{3}$$

19. Find the antiderivative $\int \frac{\sin x}{\sqrt{\cos x}}\, dx$ with the substitution method.

Solve It

20. Find the antiderivative $\int x^4 \sqrt[3]{2x^5 + 6}\, dx$ with the substitution method.

Solve It

21. Use substitution to determine $\int 5x^3 e^{x^4}\, dx$.

Solve It

22. Use substitution to antidifferentiate $\int \frac{\sec^2 \sqrt{x}}{\sqrt{x}}\, dx$.

Solve It

23. Evaluate $\int_{0}^{2} \frac{t\,dt}{\left(t^2+5\right)^4}$. Change the indices of integration.

Solve It

24. Evaluate $\int_{1}^{8} \frac{\left(s^{2/3}+5\right)^3}{\sqrt[3]{s}}\,ds$ without changing the indices of integration.

Solve It

Solutions to Reverse Differentiation Problems

1   Where (from 0 to 8) does A_g equal zero? **At about $x = 2$ or $2\frac{1}{2}$ and about $x = 6$.**

A_g equals zero twice between 0 and 8. First at about $x = 2$ or $2\frac{1}{2}$ where the negative area beginning at $x = 1$ cancels out the positive area between $x = \frac{1}{2}$ and $x = 1$. The second zero of A_g is roughly at 6 (see the dotted line in the figure). After the first zero at about 2, negative area is added between 2 and 4. The positive area from 4 to 6 roughly cancels that out, so A_g returns to zero at about 6.

2   Where (from $x = 0$ to $x = 8$) does A_g reach

     a. its maximum value? **A_g reaches its max at $x = 8$. After the zero at $x = 6$, A_g grows by about 4 or $4\frac{1}{2}$ square units by the time it gets to 8.**

     b. its minimum value? **The minimum value of A_g is at $x = 4$ where it equals something like –2.**

3   In what intervals between 0 and 8 is A_g

     a. increasing? **A_g is increasing from 0 to 1 and from 4 to 8.**

     b. decreasing? **A_g is decreasing from 1 to 4.**

4   Approximate $A_g(1)$, $A_g(3)$, and $A_g(5)$.

$A_g(1)$ is maybe a bit bigger than the right triangle with base from $x = \frac{1}{2}$ to $x = 1$ on the x-axis and vertex at $\left(\frac{1}{2}, 4\right)$. **So that area is about 1 or $1\frac{1}{4}$.**

There's a zero at about $x = 2$ or $2\frac{1}{2}$. Between there and $x = 3$ there's very roughly an area of –1, so **$A_g(3)$ is about –1.**

In problem 2b, you estimate $A_g(4)$ to be about –2. Between 4 and 5, there's sort of a triangular shape with a rough area of $\frac{1}{2}$. **Thus $A_g(5)$ equals about $-2 + \frac{1}{2}$ or $-1\frac{1}{2}$.**

5   **a.** If $A_f(x) = \int_0^x \sin t\, dt$, $\dfrac{d}{dx} A_f(x) = \sin x$.

     b. If $A_g(x) = \int_{\pi/4}^x \sin t\, dt$, $\dfrac{d}{dx} A_g(x) = \sin x$.

*6   Given that $A_f(x) = \int_{-\pi/4}^{\cos x} \sin t\, dt$, find $\dfrac{d}{dx} A_f(x)$. **The answer is $-\sin x \cdot \sin(\cos x)$.**

This is a chain rule problem. Because the derivative of $\int_{-\pi/4}^x \sin t\, dt$ is $\sin x$, the derivative of $\int_{-\pi/4}^{stuff} \sin t\, dt$ is $\sin(stuff) \cdot stuff'$. Thus the derivative of $\int_{-\pi/4}^{\cos x} \sin t\, dt$ is $\sin(\cos x) \cdot (\cos x)' = -\sin x \cdot \sin(\cos x)$.

7   For $A_f(x)$ from problem 5a, where does $\dfrac{d}{dx} A_f$ equal zero? $\dfrac{d}{dx} A_f = \sin x$, so $\dfrac{d}{dx} A_f$ is zero at **all the zeros of $\sin x$, namely at all multiples of π: $k\pi$ (for any integer, k).**

*8   For $A_f(x)$ from problem 6, evaluate $A_f'\left(\dfrac{\pi}{4}\right)$. **In problem 6, you found that**

$$A_f'(x) = -\sin x \cdot \sin(\cos x), \text{ so } A_f'\left(\frac{\pi}{4}\right) = -\sin\frac{\pi}{4} \cdot \sin\left(\cos\frac{\pi}{4}\right) = -\frac{\sqrt{2}}{2} \cdot \sin\frac{\sqrt{2}}{2} \approx -.459.$$

9   What's the area under $y = \sin x$ from 0 to π? **The area is 2.** The derivative of $-\cos x$ is $\sin x$, so $-\cos x$ is an antiderivative of $\sin x$. Thus, by the Fundamental Theorem,

$$\int_0^\pi \sin x\, dx = -\cos x \Big]_0^\pi = -(-1 - 1) = -(-2) = 2.$$

10 $\int_0^{2\pi} \sin x \, dx = -\cos x \big]_0^{2\pi} = -(1-1) = 0$ Do you see why the answer is zero?

11 $\int_2^3 (x^3 - 4x^2 + 5x - 10) \, dx = \mathbf{-6.58}.$

$$\int_2^3 (x^3 - 4x^2 + 5x - 10) \, dx$$

$$= \frac{1}{4} x^4 - \frac{4}{3} x^3 + \frac{5}{2} x^2 - 10x \Big]_2^3$$

$$= \left(\frac{1}{4} \cdot 81 - \frac{4}{3} \cdot 27 + \frac{5}{2} \cdot 9 - 30 \right) - \left(\frac{1}{4} \cdot 16 - \frac{4}{3} \cdot 8 + \frac{5}{2} \cdot 4 - 20 \right)$$

$$\approx -6.58$$

12 $\int_{-1}^2 e^x \, dx \approx \mathbf{7.02}$

$(e^x)' = e^x$, so e^x is its own antiderivative as well as its own derivative. Thus,

$$\int_{-1}^2 e^x \, dx = e^x \big]_{-1}^2 = e^2 - e^{-1} \approx 7.02.$$

13 $\int (4x - 1)^3 \, dx = \frac{\mathbf{1}}{\mathbf{16}}(\mathbf{4x - 1})^4 + \mathbf{C}$

1. Guess your answer: $\frac{1}{4}(4x - 1)^4$

2. Differentiate: $(4x - 1)^3 \cdot 4$ (by the chain rule). It's 4 times too much.

3. Tweak guess: $\frac{1}{16}(4x - 1)^4$

4. Differentiate to check: $\frac{1}{4}(4x - 1)^3 \cdot 4 = (4x - 1)^3$. Bingo.

14 $\int \sec^2 6x \, dx = \frac{\mathbf{1}}{\mathbf{6}} \tan 6x + \mathbf{C}$

Your guess at the antiderivative, $\tan 6x$, gives you $(\tan 6x)' = \sec^2 6x \cdot 6$. Tweak the guess to $\frac{1}{6} \tan 6x$. Check: $\left(\frac{1}{6} \tan 6x \right)' = \frac{1}{6} \sec^2 6x \cdot 6 = \sec^2 6x$.

15 $\int \cos \frac{x - 1}{2} \, dx = \mathbf{2 \sin} \frac{\mathbf{x - 1}}{\mathbf{2}} + \mathbf{C}$

Your guess is $\sin \frac{x - 1}{2}$. Differentiating that gives you $\cos \left(\frac{x - 1}{2} \right) \cdot \frac{1}{2}$.

Tweaked guess is $2 \sin \frac{x - 1}{2}$. That's it.

16 $\int \frac{3 \, dt}{2t + 5} = \frac{\mathbf{3}}{\mathbf{2}} \ln |2t + 5| + \mathbf{C}$

$\ln |2t + 5|$ is your guess. Differentiating gives you: $\frac{1}{2t + 5} \cdot 2$.

You wanted a 3, but you got a 2, so tweak your guess by 3 over 2. (I'm a poet!)

This "poem" always works. Try it for the other problems. Often what you want is a 1. For example, for problem 15, you'd have "You wanted a 1 but you got a ½, so tweak your guess by 1 over ½." That's 2, of course. It works!

Back to problem 16. Your tweaked guess is $\frac{3}{2} \ln |2t + 5|$. That's it.

17 $\int_0^{\pi} \frac{5}{\pi} \sec(5t - \pi) \tan(5t - \pi) \, dt = \frac{2}{\pi}$

Don't let all those 5s and πs distract you — they're just a smoke screen.

Guess: $\sec(5t - \pi)$. Diff: $\sec(5t - \pi) \tan(5t - \pi) \cdot 5$.

Tweak: $\frac{1}{\pi} \sec(5t - \pi)$. Diff: $\frac{1}{\pi} \sec(5t - \pi) \tan(5t - \pi) \cdot 5$. Bingo. So now —

$$\frac{1}{\pi} \sec(5t - \pi) \Big]_0^{\pi} = \frac{1}{\pi} \Big[\sec(4\pi) - \sec(-\pi) \Big] = \frac{2}{\pi}$$

18 $\int \frac{4.5}{1 + 9x^2} \, dx = \frac{3}{2} \tan^{-1} 3x + C$

I bet you've got the method down by now: Guess, diff, tweak, diff; Guess, diff, tweak, diff. . . .

Guess: $\tan^{-1} 3x$. Diff: $\dfrac{1}{1 + (3x)^2} \cdot 3$.

Tweak: $\frac{3}{2} \tan^{-1} 3x$. Diff: $\dfrac{3}{2} \cdot \dfrac{1}{1 + (3x)^2} \cdot 3$. That's it.

19 $\int \frac{\sin x}{\sqrt{\cos x}} \, dx = -2\sqrt{\cos x} + C$

1. It's not $\sqrt{\text{plain old } x}$, so substitute $u = \cos x$.

2. Differentiate and solve for du.

$$\frac{du}{dx} = -\sin x$$
$$du = -\sin x \, dx$$

3. Tweak inside and outside of integral: $-\int \frac{-\sin x}{\sqrt{\cos x}} \, dx$

4. Pull the switch: $= -\int \frac{du}{\sqrt{u}}$

5. Antidifferentiate with reverse power rule: $= -\int u^{-1/2} \, du = -2u^{1/2} + C$

6. Get rid of u: $-2(\cos x)^{1/2} + C = -2\sqrt{\cos x} + C$

20 $\int x^4 \sqrt[3]{2x^5 + 6} \, dx = \dfrac{3(x^5 + 3)\sqrt[3]{2x^5 + 6}}{20} + C$

1. It's not $\sqrt{\text{plain old } x}$, so substitute $u = 2x^5 + 6$.

2. Differentiate and solve for du.

$$\frac{du}{dx} = 10x^4$$
$$du = 10x^4 \, dx$$

3. Tweak inside and outside: $\frac{1}{10} \int 10x^4 \sqrt[3]{2x^5 + 6} \, dx$

4. Flip the switch: $= \frac{1}{10} \int \sqrt[3]{u} \, du$

5. Apply the power rule in reverse: $= \dfrac{1}{10} \cdot \dfrac{3}{4} u^{4/3} + C = \dfrac{3u\sqrt[3]{u}}{40} + C$

6. Switch back: $\dfrac{3(2x^5 + 6)\sqrt[3]{2x^5 + 6}}{40} + C = \dfrac{3(x^5 + 3)\sqrt[3]{2x^5 + 6}}{20} + C$

21 $\int 5x^3 e^{x^4} dx = \dfrac{5}{4} e^{x^4} + C$

1. It's not $e^{\text{plain old } x}$, so $u = x^4$.

2. You know the drill: $du = 4x^3 dx$

3. Tweak: $\dfrac{5}{4} \int 4x^3 e^{x^4} dx$

4. Switch: $= \dfrac{5}{4} \int e^u du$

5. Antidifferentiate: $= \dfrac{5}{4} e^u + C$

6. Switch back: $= \dfrac{5}{4} e^{x^4} + C$

22 $\int \dfrac{\sec^2 \sqrt{x}}{\sqrt{x}} dx = \mathbf{2 \tan \sqrt{x} + C}$

1. It's not $\sec^2(\text{plain old } x)$, so $u = \sqrt{x}$.

2. Differentiate: $du = \dfrac{1}{2} x^{-1/2} dx = \dfrac{1}{2\sqrt{x}} dx$

3. Tweak: $2 \int \dfrac{\sec^2 \sqrt{x}}{2\sqrt{x}} dx$

4. Switch: $= 2 \int \sec^2 u \, du$

5. Antidifferentiate: $= 2 \tan u + C$

6. Switch back: $= 2 \tan \sqrt{x} + C$

23 $\displaystyle\int_0^2 \dfrac{t\,dt}{\left(t^2 + 5\right)^4} \approx \mathbf{.0011}$

1. Do the *U and Diff* (it's sweeping the nation!), and find the *u*-indices of integration.

   $u = t^2 + 5 \quad$ when $t = 0$, $u = 5$

   $du = 2t\,dt \quad$ when $t = 2$, $u = 9$

2. Two tweaks: $= \dfrac{1}{2} \displaystyle\int_0^2 \dfrac{2t\,dt}{\left(t^2 + 5\right)^4}$

3. The switch: $= \dfrac{1}{2} \displaystyle\int_5^9 \dfrac{du}{u^4}$

4. Antidifferentiate and evaluate: $= \dfrac{1}{2} \cdot \left(-\dfrac{1}{3}\right) u^{-3} \Big]_5^9 = -\dfrac{1}{6}\left(9^{-3} - 5^{-3}\right) \approx .0011$

24 $\displaystyle\int_1^8 \dfrac{\left(s^{2/3} + 5\right)^3}{\sqrt[3]{s}} ds = \mathbf{1974.375}$

You know the drill: $u = s^{2/3} + 5; \quad du = \dfrac{2}{3} s^{-1/3} ds = \dfrac{2}{3\sqrt[3]{s}} ds$

$$\int_1^8 \dfrac{\left(s^{2/3} + 5\right)^3}{\sqrt[3]{s}} ds = \dfrac{3}{2} \int_1^8 \dfrac{2\left(s^{2/3} + 5\right)^3}{3\sqrt[3]{s}} ds$$

$$\dfrac{3}{2} \int_1^8 u^3 \, du$$

You'll get a math ticket if you put an equal sign in front of the last line because it is *not* equal to the line before it. When you don't change the limits of integration, you get this mixed-up integral with an integrand in terms of u, but with limits of integration in terms of x (s in this problem). This may be one reason why the preferred, book method includes switching the limits of integration — it's mathematically cleaner.

Now just antidifferentiate, switch back, and evaluate:

$$\frac{3}{2} \cdot \frac{1}{4} u^4$$

$$\frac{3}{2} \cdot \frac{1}{4} \left(s^{2/3} + 5\right)^4 \Big|_1^8 = \frac{3}{8}\left(9^4 - 6^4\right) = 1974.375$$

Chapter 11

Integration Rules for Calculus Connoisseurs

. . . .

In This Chapter

▶ Imbibing integration

▶ Transfixing on trigonometric integrals

▶ Partaking of partial fractions

. . . .

*I*n this chapter, you work on some complex and challenging integration techniques. The methods may seem quite difficult at first, but, like with anything, they're not that bad at all after some practice.

Integration by Parts: Here's How u du It

Integration by parts is the counterpart of the product rule for differentiation (see Chapter 6) because the integrand in question is the product of two functions (usually). Here's the method in a nutshell. You split up the two functions in the integrand, differentiate one, integrate the other, then apply the integration-by-parts formula. This process converts the original integrand — which you *can't* integrate — into an integrand you *can* integrate. Clear as mud, right? You'll catch on to the technique real quick if you use the following LIATE acronym and the box method in the example. First, here's the formula:

For integration by parts, here's what *u du*: $\int u\,dv = uv - \int v\,du$.

Don't try to understand that until you work through an example problem. Your first challenge in an integration by parts problem is to decide what function in your original integrand will play the role of the *u* in the formula. Here's how you do it.

To select your *u* function, just go down this list; the first function type from this list that's in your integrand is your *u*. Here's a great acronym to help you pick your *u* function: LIATE, for

  ✔ <u>L</u>ogarithmic (like ln*x*)

  ✔ <u>I</u>nverse trigonometric (like arcsin *x*)

  ✔ <u>A</u>lgebraic (like $4x^3 - 10$)

  ✔ <u>T</u>rigonometric (like sin*x*)

  ✔ <u>E</u>xponential (like 5^x)

I wish I could take credit for this acronym, but credit goes to Herbert Kasube (see his article in *American Mathematical Monthly* 90, 1983). I can, however, take credit for the following brilliant mnemonic devise to help you remember the acronym: <u>L</u>illiputians <u>I</u>n <u>A</u>frica <u>T</u>ackle <u>E</u>lephants.

Q. Integrate $\int x^2 \ln x \, dx$.

A. $\frac{1}{3} x^3 \left(\ln x - \frac{1}{3} \right) + C$

1. **Pick your *u* function.**

   The integrand contains a logarithmic function (first on the LIATE list), so ln *x* is your *u*. Everything else in the integrand — namely $x^2 \, dx$ — is automatically your *dv*.

2. **Use a box like the one in the following figure to organize the four elements of the problem.**

u	*v*
du	*dv*

   Put your *u* and your *dv* in the appropriate cells, as the following figure shows.

ln(*x*)	
	$x^2 dx$

3. **Differentiate *u* and integrate *dv*, as the arrows in the figure show.**

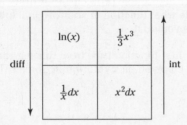

4. **Follow the arrows in the following box to help you remember how to use the integration by parts formula.**

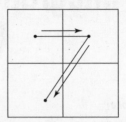

   Your original integral equals the product of the two cells along the top minus the integral of the product of the cells on the diagonal. (Think of drawing a "7" — that's your order.)

   $$\int x^2 \ln x \, dx = \ln x \cdot \frac{1}{3} x^3 - \int \left(\frac{1}{3} x^3 \cdot \frac{1}{x} \right) dx$$

5. **Simplify and integrate.**

   $$= \frac{1}{3} x^3 \ln x - \frac{1}{3} \int x^2 \, dx$$
   $$= \frac{1}{3} x^3 \ln x - \frac{1}{3} \cdot \frac{1}{3} x^3 + C$$
   $$= \frac{1}{3} x^3 \left(\ln x - \frac{1}{3} \right) + C$$

   You're done.

1. What's $\int x \cos(5x - 2)\, dx$?

Solve It

2. Integrate $\int \arctan x\, dx$. **Tip:** Sometimes integration by parts works when the integrand contains only a single function.

Solve It

3. Evaluate $\int x \arctan x\, dx$.

Solve It

4. Evaluate $\int_{-1}^{1} x 10^x\, dx$.

Solve It

*5. What's $\int x^2 e^{-x} dx$? **_Tip:_** Sometimes you have to do integration by parts more than once.

Solve It

*6. Integrate $\int e^x \sin x \, dx$. **_Tip:_** Sometimes you circle back to where you started from — that's a good thing!

Solve It

Transfiguring Trigonometric Integrals

Don't you just love trig? I'll bet you didn't realize that studying calculus was going to give you the opportunity to do so much more trig. Remember this next Thanksgiving when everyone around the dinner table is invited to mention something that they're thankful for.

This section lets you practice integrating expressions containing trigonometric functions. The basic idea is to fiddle with the integrand until you're able to finish with a u-substitution (see Chapter 10). In the next section, you use some fancy trigonometric substitutions to solve integrals that don't contain trig functions.

Q. Integrate $\int \sin^3\theta \cos^6\theta \, d\theta$.

A. $-\frac{1}{7}\cos^7\theta + \frac{1}{9}\cos^9\theta + C$

1. **Split up the $\sin^3\theta$ into $\sin^2\theta \cdot \sin\theta$ and rewrite as follows:**

   $= \int \sin^2\theta \cos^6\theta \sin\theta \, d\theta$

2. **Use the Pythagorean Identity to convert the even number of sines (the ones on the left) into cosines.**

   The Pythagorean Identity tells you that $\sin^2 x + \cos^2 x = 1$ for any angle x. If you divide both sides of this identity by

$\sin^2 x$, you get another form of the identity: $1 + \cot^2 x = \csc^2 x$. If you divide by $\cos^2 x$, you get $\tan^2 x + 1 = \sec^2 x$.

   $= \int (1 - \cos^2\theta) \cos^6\theta \sin\theta \, d\theta$

   $= \int \cos^6\theta \sin\theta \, d\theta - \int \cos^8\theta \sin\theta \, d\theta$

3. **Integrate with u-substitution with $u = \cos\theta$ for both integrals.**

   $= \int u^6(-du) - \int u^8(-du) =$

   $-\int u^6 \, du + \int u^8 \, du = -\frac{1}{7}u^7 + \frac{1}{9}u^9 + C$

   $= -\frac{1}{7}\cos^7\theta + \frac{1}{9}\cos^9\theta + C$

7. $\int \sqrt[3]{\sin x}\, \cos^3 x\, dx$

Solve It

*8. Evaluate $\int_0^{\pi/6} \cos^4 t \sin^2 t\, dt$. **_Hint:_** When the powers of both sine and cosine are even, you convert all sines and cosines into *odd* powers of cosine with these handy trig identities:

$$\sin^2 x = \frac{1 - \cos 2x}{2} \text{ and } \cos^2 x = \frac{1 + \cos 2x}{2}$$

*9. $\int \sec^3 x \tan^3 x\, dx$ **_Hints:_** 1) This works pretty much like the example in this section. 2) Convert into secants.

Solve It

*10. Evaluate $\int_{\pi/4}^{\pi/3} \tan^2 \theta \sec^4 \theta\, d\theta$. **_Hint:_** After the split-up, you convert into tangents.

Solve It

*11. $\int \tan^8 t \, dt$

Solve It

12. $\int \sqrt{\csc x} \, \cot^3 x \, dx$

Solve It

Trigonometric Substitution: It's Your Lucky Day!

In this section, you tackle integrals containing radicals of the following three forms: $\sqrt{u^2 + a^2}$, $\sqrt{u^2 - a^2}$, and $\sqrt{a^2 - u^2}$, as well as powers of these roots. To solve these problems, you use a *SohCahToa* right triangle, the Pythagorean Theorem, and some fancy trigonometric substitutions. I'm sure you'll have no trouble with this technique — it's even easier than string theory.

Q. Find $\int \dfrac{dx}{\sqrt{4x^2 + 25}}$.

A. $\dfrac{1}{2}\ln\left|\sqrt{4x^2 + 25} + 2x\right| + C$

1. **Rewrite the function to fit the form $\sqrt{u^2 + a^2}$.**

$$= \int \dfrac{dx}{\sqrt{(2x)^2 + 5^2}}$$

2. **Draw a *SohCahToa* right triangle where $\tan\theta$ equals $\dfrac{u}{a}$, namely $\dfrac{2x}{5}$.**

Note that when you make the opposite side equal to $2x$ and the adjacent side equal to 5, the hypotenuse automatically becomes your radical, $\sqrt{4x^2 + 25}$. (This follows from the Pythagorean Theorem.) See the following figure.

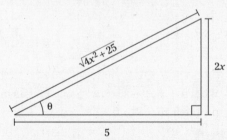

3. **Solve $\tan\theta = \dfrac{2x}{5}$ for x, differentiate, and solve for dx.**

$$\dfrac{2x}{5} = \tan\theta$$

$$x = \dfrac{5\tan\theta}{2}$$

$$\dfrac{dx}{d\theta} = \dfrac{5}{2}\sec^2\theta$$

$$dx = \dfrac{5}{2}\sec^2\theta\, d\theta$$

4. **Determine which trig function is represented by the radical over the a, then solve for the radical.**

In the figure in Step 2, the radical is on the *Hypotenuse* and the a, namely 5, is the *Adjacent* side. $\dfrac{H}{A}$ is secant so you've got

$$\dfrac{\sqrt{4x^2 + 25}}{5} = \sec\theta$$

$$\sqrt{4x^2 + 25} = 5\sec\theta$$

5. **Use the results from Steps 3 and 4 to make substitutions in the original integral and then integrate.**

$$\int \dfrac{dx}{\sqrt{4x^2 + 25}}$$

$$= \int \dfrac{\dfrac{5}{2}\sec^2\theta\, d\theta}{5\sec\theta} \quad \begin{array}{l}\leftarrow \text{from Step 3}\\ \leftarrow \text{from Step 4}\end{array}$$

$$= \dfrac{1}{2}\int \sec\theta\, d\theta$$

$$= \dfrac{1}{2}\ln|\sec\theta + \tan\theta| + C$$

Get this last integral from your textbook, the *Calculus For Dummies* cheat sheet, or from memory.

6. **Use Steps 2 and 4 or the triangle to get rid of the $\sec\theta$ and $\tan\theta$.**

$$= \dfrac{1}{2}\ln\left|\dfrac{\sqrt{4x^2 + 25}}{5} + \dfrac{2x}{5}\right| + C$$

$$= \dfrac{1}{2}\ln\left|\sqrt{4x^2 + 25} + 2x\right| - \dfrac{1}{2}\ln 5 + C$$

$$= \dfrac{1}{2}\ln\left|\sqrt{4x^2 + 25} + 2x\right| + C \quad (\dfrac{1}{2}\ln 5 + C \text{ is}$$
$$\text{just another}$$
$$\text{constant, so you}$$
$$\text{can replace it by } C.)$$

Tip: Remember that Step 2 always involves $\dfrac{u}{a}$, and Step 4 always involves $\dfrac{\sqrt{}}{a}$. How about <u>U</u> <u>A</u>re <u>R</u>adically <u>A</u>wesome?

*13. Integrate $\int x \sqrt{9x + 16x^2}\, dx$.

Solve It

14. Integrate $\int \dfrac{dx}{(9x^2 + 4)\sqrt{9x^2 + 4}}$.

Solve It

15. What's $\int \dfrac{x\, dx}{\sqrt{25 - x^2}}$? **Hint:** This is a $\sqrt{a^2 - u^2}$ problem where $\frac{u}{a} = \sin\theta$.

Solve It

16. What's $\int x \sqrt{9 - 25x^2}\, dx$? Same hint as in problem 15.

Solve It

17. Integrate $\int \dfrac{dx}{\sqrt{625x^2 - 121}}$. *Hint:* This is a $\sqrt{u^2 - a^2}$ problem where $\frac{u}{a} = \sec\theta$.

Solve It

18. Last one: $\int \dfrac{\sqrt{4x^2 - 1}}{x}\, dx$. Same hint as in problem 17.

Solve It

Partaking of Partial Fractions

The basic idea behind the partial fractions technique is what I call "unaddition" of fractions. Because $\frac{1}{2} + \frac{1}{6} = \frac{2}{3}$, had you started with $\frac{2}{3}$, you could have taken it apart — or "unadded" it — and arrived at $\frac{1}{2} + \frac{1}{6}$. You do the same thing in this section except that you do the unadding with rational functions instead of simple fractions.

Q. Integrate $\int \dfrac{3x}{x^2 - 3x - 4}\, dx$

A. $\dfrac{3}{5} \ln|x + 1| + \dfrac{12}{5} \ln|x - 4| + C$

1. Factor the denominator.

$$= \int \dfrac{3x}{(x + 1)(x - 4)}\, dx$$

2. Break up the fraction.

$$\dfrac{3x}{(x + 1)(x - 4)} = \dfrac{A}{x + 1} + \dfrac{B}{x - 4}$$

3. Multiply both sides by the denominator of the fraction on the left.

$$3x = A(x - 4) + B(x + 1)$$

4. Plug the roots of the linear factors into x one at a time.

Plug in 4: $3 \cdot 4 = B(4 + 1)$ $B = \dfrac{12}{5}$

Plug in −1: $-3 = -5A$ $A = \dfrac{3}{5}$

5. Split up the integral and integrate.

$$\int \dfrac{3x\, dx}{x^2 - 3x - 4} = \dfrac{3}{5} \int \dfrac{dx}{x + 1} + \dfrac{12}{5} \int \dfrac{dx}{x - 4}$$

$$= \dfrac{3}{5} \ln|x + 1| + \dfrac{12}{5} \ln|x - 4| + C$$

Q. Integrate $\int \dfrac{2x+1}{x^3(x^2+1)^2}\,dx$.

A. $\ln\dfrac{x^2+1}{x^2} - 3\arctan x -$

$\dfrac{2x+1}{2(x^2+1)} - \dfrac{1}{2x^2} - \dfrac{2}{x} + C$

1. **Factor the denominator. I did this step for you — a random act of kindness. Note that x^2+1 can't be factored.**

2. **Break up the fraction into a sum of fractions.**

$$\frac{2x+1}{x^3(x^2+1)^2} = \frac{A}{x} + \frac{B}{x^2} + \frac{C}{x^3} + \frac{Dx+E}{(x^2+1)} + \frac{Fx+G}{(x^2+1)^2}$$

Note the difference between the numerators of fractions with x in their denominator and those with (x^2+1) — an irreducible quadractic — in their denominator. Also note there is a fraction for each power of each different factor of the original fraction.

3. **Multiply both sides of this equation by the left-side denominator.**

$$2x+1 = Ax^2(x^2+1)^2 + Bx(x^2+1)^2 +$$
$$C(x^2+1)^2 + (Dx+E)x^3(x^2+1) +$$
$$(Fx+G)x^3$$

4. **Plug the roots of the linear factors into x (0 is the only root).**

   Plugging 0 into x eliminates every term but the "C" term. One down, six to go.

   $0 + 1 = C(0^2+1)^2$

   $C = 1$

5. **Equate coefficients of like terms.**

   Because Step 4 only gave you one term, take a different tack. If you multiply (FOIL) everything out in the Step 3 equation, the right side of the equation will contain a constant term and terms in x, x^2, x^3, x^4, x^5, and x^6. This equation is an identity, so the coefficient of, say, the x^5 term on the right has to equal the coefficient of the x^5 term on the left (which is 0 in this problem). So set the coefficient of each term on the right equal to the coefficient of its corresponding term on the left. Here's your final result:

   1) Constant term: $1 = C$

   2) x term: $2 = B$

   3) x^2 term: $0 = A + 2C$

   4) x^3 term: $0 = 2B + E + G$

5) x^4 term: $0 = 2A + C + D + F$

6) x^5 term: $0 = B + E$

7) x^6 term: $0 = A + D$

You can quickly obtain the values of all seven unknowns from these seven equations, and thus you could have skipped Step 4. But plugging in roots is so easy, and the values you get may help you finish the problem faster, so it's always a good thing to do.

And there's a third way to solve for the unknowns. You can obtain a system of equations like the one in this step by plugging *non-root* values into x. (Use small numbers that are easy to calculate with.) After doing several partial fraction problems, you'll get a feel for what combination of the three techniques works best for each problem.

From this system of equations, you get the following values:

$A = -2, B = 2, C = 1, D = 2, E = -2, F = 1, G = -2$

6. **Split up your integral and integrate.**

$$\int \frac{2x+1}{x^3(x^2+1)^2}\,dx = \int \frac{-2}{x}\,dx +$$

$$\int \frac{2}{x^2}\,dx + \int \frac{1}{x^3}\,dx + \int \frac{2x-2}{x^2+1}\,dx +$$

$$\int \frac{x-2}{(x^2+1)^2}\,dx$$

The first three are easy:

$-2\ln|x| + \dfrac{-2}{x} + \dfrac{-1}{2x^2}$. Split up the last two.

$$+\int \frac{2x}{x^2+1}\,dx - 2\int \frac{1}{x^2+1}\,dx +$$

$$\int \frac{x}{(x^2+1)^2}\,dx - 2\int \frac{1}{(x^2+1)^2}\,dx$$

The first and third above can be done with a simple u-substitution; the second is arctangent; and the fourth is very tricky, so I'm just going to give it to you:

$$+\ln(x^2+1) - 2\arctan x - \frac{1}{2(x^2+1)} -$$

$$2\left(\frac{\arctan x}{2} + \frac{x}{2(x^2+1)}\right)$$

Finally, here's the whole enchilada:

$$\int \frac{2x+1}{x^3(x^2+1)^2}\,dx = \ln\frac{x^2+1}{x^2} -$$

$$3\arctan x - \frac{2x-1}{2(x^2+1)} - \frac{1}{2x^2} - \frac{2}{x} + C$$

Take five.

19. Integrate $\int \dfrac{5dx}{2x^2 + 7x - 4}$.

Solve It

20. Integrate $\int \dfrac{2x - 3}{(3x - 1)(x + 4)(x + 5)}$.

Solve It

21. What's $\int \dfrac{x^2 + x + 1}{x^3 - 3x^2 + 3x - 1} \, dx$?

Solve It

22. Integrate $\int \dfrac{dx}{x^4 + 6x^2 + 5}$.

Solve It

*23. Integrate $\int \frac{4x^3 + 3x^2 + 2x + 1}{x^4 - 1} dx$.

Solve It

*24. What's $\int \frac{x^2 - x}{(x+1)(x^2+1)(x^2+2)} dx$?

Solve It

Solutions for Integration Rules

1 $\int x \cos(5x - 2)\, dx = \frac{1}{5} x \sin(5x - 2) + \frac{1}{25} \cos(5x - 2) + C$

1. Pick x as your u, because the algebraic function x is the first on the LIATE list.

2. Fill in your box.

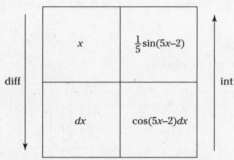

3. Use the "7" rule: $\int x \cos(5x - 2)\, dx = \frac{1}{5} x \sin(5x - 2) - \frac{1}{5} \int \sin(5x - 2)\, dx$

4. Finish by integrating: $= \frac{1}{5} x \sin(5x - 2) + \frac{1}{25} \cos(5x - 2) + C$

2 $\int \arctan x\, dx = x \arctan x - \frac{1}{2} \ln(1 + x^2) + C$

1. Pick $\arctan x$ as your u. You've got no choice.

2. Do the box thing.

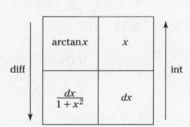

3. Apply the "7" rule: $\int \arctan x\, dx = x \arctan x - \int \frac{x\, dx}{1 + x^2} = x \arctan x - \frac{1}{2} \ln(1 + x^2) + C$

3 $\int_0^\pi x \arctan x\, dx = \frac{1}{2} x^2 \arctan x - \frac{1}{2} x + \frac{1}{2} \arctan x + C$

1. Pick $\arctan x$ as your u.

2. Do the box.

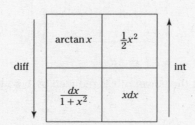

3. Apply the "7" rule.

$$\int x \arctan x \, dx = \frac{1}{2} x^2 \arctan x - \frac{1}{2} \int \frac{x^2 \, dx}{1+x^2}$$

$$= \frac{1}{2} x^2 \arctan x - \frac{1}{2} \int \frac{x^2+1-1}{1+x^2} \, dx$$

$$= \frac{1}{2} x^2 \arctan x - \frac{1}{2} \int dx + \frac{1}{2} \int \frac{dx}{1+x^2}$$

$$= \frac{1}{2} x^2 \arctan x - \frac{1}{2} x + \frac{1}{2} \arctan x + C \quad \text{or} \quad \frac{x^2+1}{2} \arctan x - \frac{x}{2} + C$$

4 $\displaystyle\int_{-1}^{1} x10^x \, dx = \frac{101 \ln 10 - 99}{10 \left(\ln 10 \right)^2}$

1. Pick the algebraic x as your u.

2. Box it.

x	$\dfrac{10^x}{\ln 10}$
dx	$10^x dx$

diff ↓   int ↑

3. Do the "7".

$$\int_{-1}^{1} x10^x \, dx = \frac{x10^x}{\ln 10} \Bigg]_{-1}^{1} - \frac{1}{\ln 10} \int_{-1}^{1} 10^x \, dx$$

$$= \frac{10}{\ln 10} + \frac{1}{10 \ln 10} - \frac{1}{\ln 10} \cdot \frac{10^x}{\ln 10} \Bigg]_{-1}^{1}$$

$$= \frac{10}{\ln 10} + \frac{1}{10 \ln 10} - \frac{10}{\left(\ln 10 \right)^2} + \frac{1}{10 \left(\ln 10 \right)^2}$$

$$= \frac{101 \ln 10 - 99}{10 \left(\ln 10 \right)^2}$$

* 5 $\displaystyle\int x^2 e^{-x} \, dx = -e^{-x} \left(x^2 + 2x + 2 \right) + C$

1. Pick x^2 as your u.

2. Box it.

x^2	$-e^{-x}$
$2x dx$	$e^{-x} dx$

diff ↓   int ↑

3. "7" it: $\displaystyle\int x^2 e^{-x} \, dx = -x^2 e^{-x} + 2 \int xe^{-x} \, dx$

In the second integral, the power of x is reduced by 1, so you're making progress.

4. Repeat the process for the second integral: Pick it and box it.

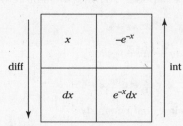

5. "7" rule for the second integral: $\int xe^{-x}\,dx = -xe^{-x} + \int e^{-x}\,dx = -xe^{-x} - e^{-x} + C$

6. Take this result and plug it into the second integral from Step 3.

$$\int x^2 e^{-x}\,dx = -x^2 e^{-x} + 2\left(-xe^{-x} - e^{-x} + C\right)$$
$$= -x^2 e^{-x} - 2xe^{-x} - 2e^{-x} + C = -e^{-x}\left(x^2 + 2x + 2\right) + C$$

*6 $\int e^x \sin x\,dx = \dfrac{e^x \sin x}{2} - \dfrac{e^x \cos x}{2} + C$

1. Pick $\sin x$ as your u — it's a T from LIATE.

2. Box it.

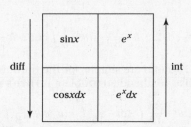

3. "7" it: $\int e^x \sin x\,dx = e^x \sin x - \int e^x \cos x\,dx$

   Doesn't look like progress, but it is. Repeat this process for $\int e^x \cos x\,dx$.

4. Pick $\cos x$ as your u and box it.

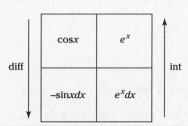

5. "7" it: $\int e^x \cos x\,dx = e^x \cos x + \int e^x \sin x\,dx$

   The prodigal son returns home and is rewarded.

6. Plug this result into the second integral from Step 3.

$$\int e^x \sin x\,dx = e^x \sin x - e^x \cos x - \int e^x \sin x\,dx$$

7. You want to solve for $\int e^x \sin x \, dx$, so bring them both to the left side and solve.

$$2 \int e^x \sin x \, dx = e^x \sin x - e^x \cos x + C$$

$$\int e^x \sin x \, dx = \frac{e^x \sin x}{2} - \frac{e^x \cos x}{2} + C$$

7 $\quad \int \sqrt[3]{\sin x} \cos^3 x \, dx = \frac{3}{4} \sin^{4/3} x - \frac{3}{10} \sin^{10/3} x + C$

1. Split off one $\cos x$: $\int \sqrt[3]{\sin x} \cos^2 x \cos x \, dx$

2. Convert the even number of cosines into sines with the Pythagorean Identity.

$$= \int \sqrt[3]{\sin x}\,(1 - \sin^2 x)\cos x \, dx = \int \sqrt[3]{\sin x}\,\cos x \, dx - \int \sin^{7/3} x \cos x \, dx$$

3. Integrate with u-substitution with $u = \sin x$: $= \frac{3}{4} \sin^{4/3} x - \frac{3}{10} \sin^{10/3} x + C$

* 8 $\quad \displaystyle\int_0^{\pi/6} \cos^4 t \sin^2 t \, dt = \frac{\pi}{96}$

1. Convert to odd powers of cosine with trig identities $\cos^2 x = \dfrac{1 + \cos 2x}{2}$ and $\sin^2 x = \dfrac{1 - \cos 2x}{2}$.

$$= \int_0^{\pi/6} \left(\frac{1 + \cos 2t}{2}\right)^2 \frac{1 - \cos 2t}{2} \, dt$$

2. Simplify and FOIL.

$$= \frac{1}{8}\int_0^{\pi/6} (1 - \cos^2 2t)(1 + \cos 2t)\, dt = \frac{1}{8}\int_0^{\pi/6} 1\,dt + \frac{1}{8}\int_0^{\pi/6} \cos 2t\,dt - \frac{1}{8}\int_0^{\pi/6} \cos^2 2t\,dt - \frac{1}{8}\int_0^{\pi/6} \cos^3 2t\,dt$$

3. Integrate. The first and second are simple; for the third, you use the same trig identity again; the fourth is handled like you handled problem 7. Here's what you should get:

$$= \frac{1}{8}\int_0^{\pi/6} 1\,dt + \frac{1}{8}\int_0^{\pi/6} \cos 2t\,dt - \frac{1}{16}\int_0^{\pi/6} 1\,dt - \frac{1}{16}\int_0^{\pi/6} \cos 4t\,dt - \frac{1}{8}\int_0^{\pi/6} \cos 2t\,dt + \frac{1}{8}\int_0^{\pi/6} \sin^2 2t \cos 2t\,dt$$

$$= \frac{1}{16}\int_0^{\pi/6} dt - \frac{1}{16}\int_0^{\pi/6} \cos 4t\,dt + \frac{1}{8}\int_0^{\pi/6} \sin^2 2t \cos 2t\,dt$$

$$= \frac{1}{16} t\,\Big]_0^{\pi/6} - \frac{1}{64} \sin 4t\,\Big]_0^{\pi/6} + \frac{1}{48} \sin^3 2t\,\Big]_0^{\pi/6}$$

$$= \frac{\pi}{96} - \frac{\sqrt{3}}{128} + \frac{\sqrt{3}}{128}$$

$$= \frac{\pi}{96}$$

* 9 $\quad \int \sec^3 x \tan^3 x \, dx = \frac{1}{5} \sec^5 x - \frac{1}{3} \sec^3 x + C$

1. Split off $\sec x \tan x$: $= \int \sec^2 x \tan^2 x \sec x \tan x \, dx$

2. Use the Pythagorean Identity to convert the even number of tangents into secants.

$$= \int \sec^2 x\,(\sec^2 x - 1) \sec x \tan x \, dx$$

$$= \int \sec^4 x \sec x \tan x \, dx - \int \sec^2 x \sec x \tan x \, dx$$

3. Integrate with u-substitution: $= \frac{1}{5} \sec^5 x - \frac{1}{3} \sec^3 x + C$

* 10 $\displaystyle\int_{\pi/4}^{\pi/3} \tan^2\theta \sec^4\theta\, d\theta = \dfrac{14\sqrt{3}}{5} - \dfrac{8}{15}$

1. Split off a $\sec^2\theta$: $= \displaystyle\int_{\pi/4}^{\pi/3} \tan^2\theta \sec^2\theta \sec^2\theta\, d\theta$

2. Convert to tangents: $= \displaystyle\int_{\pi/4}^{\pi/3} \tan^2\theta\left(\tan^2\theta + 1\right)\sec^2\theta\, d\theta = \int_{\pi/4}^{\pi/3} \tan^4\theta \sec^2\theta\, d\theta + \int_{\pi/4}^{\pi/3} \tan^2\theta \sec^2\theta\, d\theta$

3. Do u-substitution with $u = \tan\theta$.

$$= \frac{1}{5}\tan^5\theta\bigg]_{\pi/4}^{\pi/3} + \frac{1}{3}\tan^3\theta\bigg]_{\pi/4}^{\pi/3}$$

$$= \frac{1}{5}\sqrt{3}^5 - \frac{1}{5}\cdot 1^5 + \frac{1}{3}\sqrt{3}^3 - \frac{1}{3}\cdot 1^3$$

$$= \frac{14\sqrt{3}}{5} - \frac{8}{15}$$

* 11 $\displaystyle\int \tan^8 t\, dt = \frac{1}{7}\tan^7 t - \frac{1}{5}\tan^5 t + \frac{1}{3}\tan^3 t - \tan t + t + C$

1. Split off a $\tan^2 t$ and convert it to secants:

$$= \int \tan^6 t\left(\sec^2 t - 1\right)dt = \int\left(\tan^6 t \sec^2 t\right)dt - \int\left(\tan^6 t\right)dt$$

2. Do the first integral with a u-substitution and repeat Step 1 with the second, then keep repeating until you get rid of all the tangents in the second integral.

$$= \frac{1}{7}\tan^7 t - \int \tan^4 t\left(\sec^2 t - 1\right)dt$$

$$= \frac{1}{7}\tan^7 t - \frac{1}{5}\tan^5 t + \int \tan^2 t\left(\sec^2 t - 1\right)dt$$

$$= \frac{1}{7}\tan^7 t - \frac{1}{5}\tan^5 t + \frac{1}{3}\tan^3 t - \int\left(\sec^2 t - 1\right)dt$$

$$= \frac{1}{7}\tan^7 t - \frac{1}{5}\tan^5 t + \frac{1}{3}\tan^3 t - \tan t + t + C$$

12 $\displaystyle\int \sqrt{\csc x}\,\cot^3 x\, dx = -\frac{2}{5}\csc^{5/2} x + 2\csc^{1/2} x + C$

1. Split off $\csc x \cot x$: $= \displaystyle\int \csc^{-1/2} x \cot^2 x \csc x \cot x\, dx$.

2. Convert the even number of cotangents to cosecants with the Pythagorean Identity.

$$= \int \csc^{-1/2} x\left(\csc^2 x - 1\right)\csc x \cot x\, dx$$

3. Finish with a u-substitution.

$$= \int \csc^{3/2} x \csc x \cot x\, dx - \int \csc^{-1/2} x \csc x \cot x\, dx$$

$$= \int u^{3/2}(-du) - \int u^{-1/2}(-du)$$

$$= -\frac{2}{5}u^{5/2} + 2u^{1/2} + C$$

$$= -\frac{2}{5}\csc^{5/2} x + 2\csc^{1/2} x + C$$

* 13 $\displaystyle\int x\sqrt{9 + 16x^2}\, dx = \frac{1}{48}\left(9 + 16x^2\right)^{3/2} + C$

1. Rewrite as $\displaystyle\int x\sqrt{(4x)^2 + 3^2}\, dx$.

2. Draw your *SohCahToa* triangle where $\tan\theta = \dfrac{u}{a}$. See the following figure.

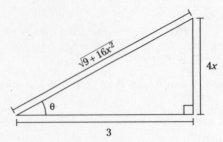

3. Solve $\tan\theta = \frac{4x}{3}$ for x, differentiate, and solve for dx.

$$4x = 3\tan\theta \qquad \frac{dx}{d\theta} = \frac{3}{4}\sec^2\theta$$

$$x = \frac{3}{4}\tan\theta \qquad dx = \frac{3}{4}\sec^2\theta\, d\theta$$

4. Do the $\frac{\sqrt{\ }}{a}$ thing.

$$\frac{\sqrt{9+16x^2}}{3} = \sec\theta \qquad \sqrt{9+16x^2} = 3\sec\theta$$

5. Substitute. **Hint:** There are three substitutions here, not just two like in the example.

$$\int x\sqrt{9+16x^2}\,dx$$

$$= \int \left(\frac{3}{4}\tan\theta\right)(3\sec\theta)\left(\frac{3}{4}\sec^2\theta\, d\theta\right)$$

$$= \frac{27}{16}\int \tan\theta\sec^3\theta\, d\theta$$

6. Now you're back in trigonometric integral territory. Split off a $\sec\theta\tan\theta$ factor.

$$= \frac{27}{16}\int \sec^2\theta\,(\sec\theta\tan\theta)\, d\theta$$

7. Integrate. This is in $\int u^2\,du$ form, so $= \frac{27}{16}\left(\frac{1}{3}\sec^3\theta\right) + C = \frac{9}{16}\sec^3\theta + C$.

8. Switch back to x: $= \frac{9}{16}\left(\frac{\sqrt{9+16x^2}}{3}\right)^3 + C = \frac{1}{48}\left(9+16x^2\right)^{3/2} + C$.

14 $\displaystyle\int \frac{dx}{(9x^2+4)\sqrt{9x^2+4}} = \frac{x}{4\sqrt{9x^2+4}} + C.$

1. Rewrite as $\displaystyle\int \frac{dx}{\sqrt{(3x)^2+2^2}^{\,3}}$.

2. Draw your triangle, remembering that $\tan\theta = \frac{u}{a}$. See the following figure.

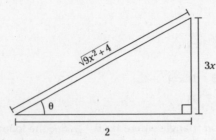

3. Solve $\tan\theta = \frac{3x}{2}$ for x, differentiate, and solve for dx.

$$3x = 2\tan\theta \qquad x = \frac{2}{3}\tan\theta \qquad dx = \frac{2}{3}\sec^2\theta\, d\theta$$

4. Do the $\frac{\sqrt{\ }}{a}$ thing.

$$\frac{\sqrt{9x^2+4}}{2} = \sec\theta \qquad \sqrt{9x^2+4} = 2\sec\theta$$

5. Substitute.

$$\int \frac{dx}{\sqrt{9x^2+4}^{\,3}}$$

$$= \int \frac{\frac{2}{3}\sec^2\theta\,d\theta}{(2\sec\theta)^3} = \frac{1}{12}\int \frac{d\theta}{\sec\theta} = \frac{1}{12}\int \cos\theta\,d\theta$$

6. Integrate to get $= \frac{1}{12}\sin\theta + C$.

7. Switch back to x (use the triangle).

$$= \frac{1}{12}\left(\frac{3x}{\sqrt{9x^2+4}}\right) + C = \frac{x}{4\sqrt{9x^2+4}} + C$$

15 $\int \dfrac{x\,dx}{\sqrt{25-x^2}} = -\sqrt{25-x^2} + C$

1. Rewrite as $\int \dfrac{x\,dx}{\sqrt{5^2-x^2}}$.

2. Draw your triangle. For this problem, $\sin\theta = \frac{u}{a}$. See the following figure.

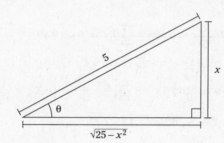

3. Solve $\sin\theta = \frac{x}{5}$ for x, and then get dx.

$$x = 5\sin\theta \qquad dx = 5\cos\theta\,d\theta$$

4. Do the $\frac{\sqrt{\ }}{a}$ thing.

$$\frac{\sqrt{25-x^2}}{5} = \cos\theta \qquad \sqrt{25-x^2} = 5\cos\theta$$

5. Substitute.

$$\int \frac{x\,dx}{\sqrt{25-x^2}}$$

$$= \int \frac{(5\sin\theta)(5\cos\theta\,d\theta)}{5\cos\theta} = 5\int \sin\theta\,d\theta$$

6. Integrate to get $-5\cos\theta + C$.

7. Switch back to x (look at Step 4): $= -5\left(\frac{\sqrt{25-x^2}}{5}\right) + C = -\sqrt{25-x^2} + C$

16 $\int x \sqrt{9 - 25x^2} \, dx = -\dfrac{1}{75} \left(9 - 25x^2\right)^{3/2} + C$

1. Rewrite as $\int x \sqrt{3^2 - (5x)^2} \, dx$.

2. Do the triangle thing. See the following figure.

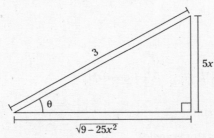

3. Solve $\sin\theta = \dfrac{5x}{3}$ for x then get dx.

$$x = \frac{3}{5} \sin\theta \qquad dx = \frac{3}{5} \cos\theta \, d\theta$$

4. Do the <u>R</u>adically <u>A</u>wesome thing.

$$\frac{\sqrt{9 - 25x^2}}{3} = \cos\theta \qquad \sqrt{9 - 25x^2} = 3\cos\theta$$

5. Substitute.

$$\int x \sqrt{9 - 25x^2} \, dx$$

$$= \int \left(\frac{3}{5} \sin\theta\right)(3\cos\theta)\left(\frac{3}{5}\cos\theta \, d\theta\right) = \frac{27}{25} \int \cos^2\theta \sin\theta \, d\theta$$

6. Integrate: $= \dfrac{27}{25}\left(-\dfrac{1}{3}\cos^3\theta\right) + C = -\dfrac{9}{25}\cos^3\theta + C$.

7. Switch back to x (look at Step 4): $= -\dfrac{9}{25}\left(\dfrac{\sqrt{9 - 25x^2}}{3}\right)^3 + C = -\dfrac{1}{75}\left(9 - 25x^2\right)^{3/2} + C$.

17 $\int \dfrac{dx}{\sqrt{625x^2 - 121}} = \dfrac{1}{25} \ln\left|25x + \sqrt{625x^2 - 121}\right| + C$

1. Rewrite as $\int \dfrac{dx}{\sqrt{(25x)^2 - 11^2}}$.

2. Do the triangle thing. For this problem, $\sec\theta = \dfrac{u}{a}$. See the following figure.

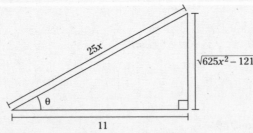

3. Solve $\sec\theta = \dfrac{25x}{11}$ for x and find dx.

$$x = \frac{11}{25} \sec\theta \qquad dx = \frac{11}{25} \sec\theta \tan\theta \, d\theta$$

4. The $\frac{\sqrt{}}{a}$ thing.

$$\frac{\sqrt{625x^2-121}}{11}=\tan\theta \qquad \sqrt{625x^2-121}=11\tan\theta$$

5. Substitute.

$$\int \frac{dx}{\sqrt{625x^2-121}}$$

$$=\int \frac{\frac{11}{25}\sec\theta\tan\theta\,d\theta}{11\tan\theta}=\frac{1}{25}\int \sec\theta\,d\theta$$

6. Integrate: $=\frac{1}{25}\ln|\sec\theta+\tan\theta|+C$.

7. Switch back to x (look at Steps 3 and 4):

$$=\frac{1}{25}\ln\left|\frac{25x}{11}+\frac{\sqrt{625x^2-121}}{11}\right|+C$$

$$=\frac{1}{25}\ln\left|25x+\sqrt{625x^2-121}\right|-\frac{1}{25}\ln 11+C$$

$$=\frac{1}{25}\ln\left|25x+\sqrt{625x^2-121}\right|+C$$

18 $\int \frac{\sqrt{4x^2-1}}{x}\,dx=\sqrt{4x^2-1}-\arctan\sqrt{4x^2-1}+C$

1. Rewrite as $=\int \frac{\sqrt{(2x)^2-1^2}}{x}\,dx$.

2. Draw your triangle. See the following figure.

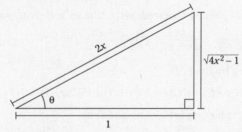

3. Solve $\sec\theta=\frac{2x}{1}$ for x; get dx.

$$x=\frac{1}{2}\sec\theta \qquad dx=\frac{1}{2}\sec\theta\tan\theta\,d\theta$$

4. Do the $\frac{\sqrt{}}{a}$ thing.

$$\sqrt{4x^2-1}=\tan\theta$$

5. Substitute.

$$\int \frac{\sqrt{4x^2-1}}{x}\,dx$$

$$=\int \frac{\tan\theta}{\frac{1}{2}\sec\theta}\cdot\frac{1}{2}\sec\theta\tan\theta\,d\theta=\int \tan^2\theta\,d\theta$$

6. Integrate: $= \int \left(\sec^2 \theta - 1 \right) d\theta = \tan \theta - \theta + C$.

7. Switch back to x (look at Step 4).

$$= \sqrt{4x^2 - 1} - \arctan \sqrt{4x^2 - 1} + C$$

or

$$= \sqrt{4x^2 - 1} - \operatorname{arcsec} 2x + C$$

 $\int \dfrac{5dx}{2x^2 + 7x - 4} = \dfrac{5}{9} \ln \left| \dfrac{2x-1}{x+4} \right| + C$

1. Factor the denominator: $= \int \dfrac{5dx}{(2x-1)(x+4)}$.

2. Break up the fraction into a sum of partial fractions: $\dfrac{5}{(2x-1)(x+4)} = \dfrac{A}{2x-1} + \dfrac{B}{x+4}$.

3. Multiply both sides by the least common denominator: $5 = A(x+4) + B(2x-1)$.

4. Plug the roots of the factors into x one at a time.

$x = -4$ gives you $x = \dfrac{1}{2}$ gives you

$5 = -9B$ $5 = \dfrac{9}{2} A$

$B = -\dfrac{5}{9}$ $A = \dfrac{10}{9}$

5. Split up your integral and integrate.

$$\int \frac{5dx}{2x^2 + 7x - 4} = \frac{10}{9} \int \frac{dx}{2x-1} + \frac{-5}{9} \int \frac{dx}{x+4} = \frac{5}{9} \ln|2x-1| - \frac{5}{9} \ln|x+4| + C = \frac{5}{9} \ln \left| \frac{2x-1}{x+4} \right| + C$$

20 $\int \dfrac{2x-3}{(3x-1)(x+4)(x+5)} = \dfrac{-7}{208} \ln|3x-1| + \dfrac{11}{13} \ln|x+4| - \dfrac{13}{16} \ln|x+5| + C$.

1. The denominator is already factored, so go ahead and write your sum of partial fractions.

$$\frac{2x-3}{(3x-1)(x+4)(x+5)} = \frac{A}{3x-1} + \frac{B}{x+4} + \frac{C}{x+5}$$

2. Multiply both sides by the LCD.

$$2x - 3 = A(x+4)(x+5) + B(3x-1)(x+5) + C(3x-1)(x+4)$$

3. Plug the roots of the factors into x one at a time.

$x = \dfrac{1}{3}$ gives you: $\quad -\dfrac{7}{3} = \dfrac{208}{9} A; \quad A = -\dfrac{21}{208}$

$x = -4 \quad " \quad " \quad ": \quad -11 = -13B; \quad B = \dfrac{11}{13}$

$x = -5 \quad " \quad " \quad ": \quad -13 = 16C \quad \quad C = -\dfrac{13}{16}$

4. Split up and integrate.

$$\int \frac{2x-3}{(3x-1)(x+4)(x+5)} dx = \frac{-21}{208} \int \frac{dx}{3x-1} + \frac{11}{13} \int \frac{dx}{x+4} + \frac{-13}{16} \int \frac{dx}{x+5}$$

$$= \frac{-7}{208} \ln|3x-1| + \frac{11}{13} \ln|x+4| - \frac{13}{16} \ln|x+5| + C$$

21 $\displaystyle\int \frac{x^2+x+1}{x^3-3x^2+3x-1}\,dx = \ln|x-1| - \frac{3(2x-1)}{2(x-1)^2} + C$

1. Factor the denominator: $= \displaystyle\int \frac{x^2+x+1}{(x-1)^3}\,dx$.

2. Write the partial fractions: $\dfrac{x^2+x+1}{(x-1)^3} = \dfrac{A}{x-1} + \dfrac{B}{(x-1)^2} + \dfrac{C}{(x-1)^3}$.

3. Multiply by the LCD: $x^2+x+1 = A(x-1)^2 + B(x-1) + C$.

4. Plug in the single root, which is 1, giving you $C = 3$.

5. Equate coefficients of like terms.

   Without multiplying out the entire right side in Step 3, you can see that the x^2 term on the right will be Ax^2. Because the coefficient of x^2 on the left is 1, A must equal 1.

6. Plug in 0 for x, giving you $1 = A - B + C$.

   Because you know A is 1 and C is 3, B must be 3.

   Note: You can solve for A, B, and C in many ways, but the way I did it is probably the quickest.

7. Split up and integrate.

$$\int \frac{x^2+x+1}{x^3-3x^2+3x-1}\,dx = \int \frac{dx}{x-1} + 3\int \frac{dx}{(x-1)^2} + 3\int \frac{dx}{(x-1)^3} = \ln|x-1| - \frac{3}{x-1} - \frac{3}{2(x-1)^2} + C$$

22 $\displaystyle\int \frac{dx}{x^4+6x^2+5} = \frac{1}{4}\arctan x - \frac{\sqrt{5}}{20}\arctan \frac{x\sqrt{5}}{5} + C$

1. Factor: $= \displaystyle\int \frac{dx}{(x^2+5)(x^2+1)}$.

2. Write the partial fractions: $\dfrac{1}{(x^2+5)(x^2+1)} = \dfrac{Ax+B}{x^2+5} + \dfrac{Cx+D}{x^2+1}$.

3. Multiply by LCD: $1 = (Ax+B)(x^2+1) + (Cx+D)(x^2+5)$.

4. Plug in the easiest numbers to work with, 0 and 1, to effortlessly get two equations.

   $x = 0:\quad 1 = B + 5D$

   $x = 1:\quad 1 = 2A + 2B + 6C + 6D$

5. After FOILing out the equation in Step 3, equate coefficients of like terms to come up with two more equations.

   The x^2 term gives you $0 = B + D$

   This equation plus the first one in Step 4 give you $B = -\dfrac{1}{4}$, $D = \dfrac{1}{4}$

   The x^3 term gives you $0 = A + C$

   Now this equation plus the second one in Step 4 plus the known values of B and D give you $A = 0$ and $C = 0$.

6. Split up and integrate.

$$\int \frac{dx}{x^4+6x^2+5} = \int \frac{-\frac{1}{4}\,dx}{x^2+5} + \int \frac{\frac{1}{4}\,dx}{x^2+1}$$

$$= -\frac{1}{4}\int \frac{dx}{x^2+5} + \frac{1}{4}\int \frac{dx}{x^2+1}$$

$$= -\frac{1}{4\sqrt{5}}\arctan \frac{x}{\sqrt{5}} + \frac{1}{4}\arctan x + C$$

★ 23 $\int \frac{4x^3 + 3x^2 + 2x + 1}{x^4 - 1} \, dx = \frac{1}{2} \ln\left[(x^2+1)|x-1|^5|x+1|\right] + \arctan x + C$

1. Factor: $= \int \frac{4x^3 + 3x^2 + 2x + 1}{(x-1)(x+1)(x^2+1)} \, dx$.

2. Write the partial fractions: $\frac{4x^3 + 3x^2 + 2x + 1}{(x-1)(x+1)(x^2+1)} = \frac{A}{x-1} + \frac{B}{x+1} + \frac{Cx+D}{x^2+1}$.

3. Multiply by the LCD:

$$4x^3 + 3x^2 + 2x + 1 = A(x+1)(x^2+1) + B(x-1)(x^2+1) + (Cx+D)(x-1)(x+1)$$

4. Plug in roots.

$\quad x = 1: \quad\quad 10 = 4A \quad\quad A = 2.5$

$\quad x = -1: \quad -2 = -4B \quad B = 0.5$

5. Equating the coefficients of the x^3 term gives you C.

$\quad 4 = A + B + C$

$\quad A = 2.5, \ B = .5, \ \text{so} \ C = 1$

6. Plugging in zero and the known values of A, B, and C gets you D.

$\quad 1 = 2.5 - 0.5 - D$

$\quad D = 1$

7. Integrate.

$$\int \frac{4x^3 + 3x^2 + 2x + 1}{x^4 - 1} \, dx = 2.5 \int \frac{dx}{x-1} + .5 \int \frac{dx}{x+1} + \int \frac{x+1}{x^2+1} \, dx$$
$$= 2.5 \ln|x-1| + .5 \ln|x+1| + .5 \ln|x^2+1| + \arctan x + C$$
$$= \frac{1}{2} \ln\left[(x^2+1)|x-1|^5|x+1|\right] + \arctan x + C$$

★ 24 $\int \frac{x^2-x}{(x+1)(x^2+1)(x^2+2)} \, dx = \frac{1}{6} \ln \frac{(x+1)^2}{x^2+2} - \arctan x + \frac{2\sqrt{2}}{3} \arctan \frac{x\sqrt{2}}{2} + C$

1. Break the already factored function into partial fractions.

$$\frac{x^2-x}{(x+1)(x^2+1)(x^2+2)} = \frac{A}{x+1} + \frac{Bx+C}{x^2+1} + \frac{Dx+E}{x^2+2}$$

2. Multiply by LCD.

$$x^2 - x = A(x^2+1)(x^2+2) + (Bx+C)(x+1)(x^2+2) + (Dx+E)(x+1)(x^2+1)$$

3. Plug in single root (–1).

$\quad 2 = 6A \quad\quad A = \frac{1}{3}$

4. Plug 0, 1, and –2 into x and $\frac{1}{3}$ into A.

$\quad x = 0: \quad\quad 0 = \frac{2}{3} + 2C + E$

$\quad x = 1: \quad\quad 0 = 2 + 6B + 6C + 4D + 4E$

$\quad x = -2: \quad\ 6 = 10 + 12B - 6C + 10D - 5E$

5. Equate coefficients of the x^4 terms (with $A = \frac{1}{3}$): $0 = \frac{1}{3} + B + D$.

6. Solve the system of four equations from Steps 4 and 5. You get the following:

$$B = 0 \qquad C = -1 \qquad D = -\frac{1}{3} \qquad E = \frac{4}{3}$$

If you find an easier way to solve for A through E, go to my Web site and send me an e-mail.

7. Integrate.

$$\int \frac{x^2 - x}{(x+1)(x^2+1)(x^2+2)} = \frac{1}{3} \int \frac{dx}{x+1} - \int \frac{dx}{x^2+1} - \frac{1}{3} \int \frac{x-4}{x^2+2} \, dx$$

$$= \frac{1}{3} \ln|x+1| - \arctan x - \frac{1}{6} \ln(x^2+2) + \frac{2\sqrt{2}}{3} \arctan \frac{x\sqrt{2}}{2} + C$$

$$= \frac{1}{6} \ln \frac{(x+1)^2}{x^2+2} - \arctan x + \frac{2\sqrt{2}}{3} \arctan \frac{x\sqrt{2}}{2} + C$$

Chapter 12

Who Needs Freud? Using the Integral to Solve Your Problems

In This Chapter

▶ Weird areas, surfaces, and volumes
▶ L'Hôpital's Rule
▶ Misbehaving integrals
▶ Other stuff you'll never use

Now that you're an expert at integrating, it's time to put that awesome power to use to solve some . . . *ahem* . . . real-world problems. All right, I admit it — the problems you see in this chapter won't seem to bear much connection to reality. But, in fact, integration is a powerful and practical mathematical tool. Engineers, scientists, and economists, among others, do important, practical work with integration that they couldn't do without it.

Finding a Function's Average Value

With differentiation, you can determine the maximum and minimum heights of a function, its steepest points, its inflection points, its concavity, and so on. But there's a simple question about a function that differentiation cannot answer: What's the function's average height? To answer that, you need integration.

Q. What's the average value of $\sin x$ between 0 and π?

A. Piece o' cake. This is a one-step problem:

$$\begin{aligned} average\ value = \frac{total\ area}{base} &= \frac{\int_{0}^{\pi} \sin x\, dx}{\pi - 0} \\ &= \frac{-\cos x \big]_{0}^{\pi}}{\pi} \\ &= \frac{-1(-1-1)}{\pi} \\ &= \frac{2}{\pi} \end{aligned}$$

1. What's the average value of $f(x) = \dfrac{x}{(x^2+1)^3}$ from 1 to 3?

Solve It

2. A car's speed in feet per second is given by $f(t) = t^{1.7} - 6t + 80$. What's its average speed from $t = 5$ seconds to $t = 15$ seconds? What's that in miles per hour?

Solve It

Finding the Area between Curves

In elementary school and high school geometry, you learned area formulas for all sorts of shapes like rectangles, circles, triangles, parallelograms, kites, and so on. Big deal. With integration, you can determine things like the area between $f(x) = x^2$ and $g(x) = \arctan x$ — now that *is* something.

Q. What's the area between $\sin x$ and $\cos x$ from 0 to π?

A. The area is $2\sqrt{2}$.

  1. Graph the two functions to get a feel for the size of the area in question and where the functions intersect.

  2. Find the point of intersection.

  $$\sin x = \cos x$$
  $$\frac{\sin x}{\cos x} = 1$$
  $$\tan x = 1$$
  $$x = \frac{\pi}{4}$$

  3. Figure the area from 0 to $\frac{\pi}{4}$.

  Between 0 and $\frac{\pi}{4}$, cosine is on top so you want cosine minus sine:

$$Area = \int_0^{\pi/4} (\cos x - \sin x)\, dx$$

$$= \sin x + \cos x \Big]_0^{\pi/4}$$

$$= \frac{\sqrt{2}}{2} + \frac{\sqrt{2}}{2} - (0 + 1)$$

$$= \sqrt{2} - 1$$

  4. Figure the area between $\frac{\pi}{4}$ and π.

  This time sine's on top:

$$Area = \int_{\pi/4}^{\pi} (\sin x - \cos x)\, dx$$

$$= -\cos x - \sin x \Big]_{\pi/4}^{\pi}$$

$$= -(-1) - 0 - \left(-\frac{\sqrt{2}}{2} - \frac{\sqrt{2}}{2} \right)$$

$$= 1 + \sqrt{2}$$

  5. Add the two areas for your final answer.

$$\sqrt{2} - 1 + 1 + \sqrt{2} = 2\sqrt{2}$$

3. What's the area enclosed by $f(x) = x^2$ and $g(x) = \sqrt{x}$?

Solve It

4. What's the total area enclosed by $f(t) = t^3$ and $g(t) = t^5$?

Solve It

*5. The lines $y = x$, $y = 2x - 5$, and $y = -2x + 3$ form a triangle in the first and fourth quadrants. What's the area of this triangle?

Solve It

6. What's the area of the triangular shape in the first quadrant between roughly x= .5 and x= 1 enclosed by $\sin x$, $\cos x$, and the line $y = \frac{1}{2}$?

Solve It

Volumes of Weird Solids: No, You're Never Going to Need This

Integration works by cutting something up into an infinite number of infinitesimal pieces and then adding the pieces up to compute the total. In this way, integration is able to determine the volume of bizarre shapes: It cuts the shapes up into thin pieces that have ordinary shapes which can then be handled by ordinary geometry. This section shows you three different methods:

✔ **The meat slicer method:** This works just like a deli meat slicer — you cut a shape into flat, thin slices. You then add up the volume of the slices. This method is used for odd, sometimes asymmetrical shapes.

✔ **The disk/washer method:** With this method, you cut up the given shape into thin, flat disks or washers (you know — like pancakes or squashed donuts). Used for shapes with circular cross-sections.

✔ **The cylindrical shell method:** Here, you cut your volume up into thin nested shells. Each one fits snugly inside the next widest one, like telescoping tubes or nested Russian dolls. Also used for shapes with circular cross-sections.

Q. What's the volume of the shape shown in the following figure? Its base is formed by the functions $f(x) = \sqrt{x}$ and $g(x) = -\sqrt{x}$. Its cross-sections are isosceles triangles whose heights grow linearly from zero at the origin to 1 when $x = 1$.

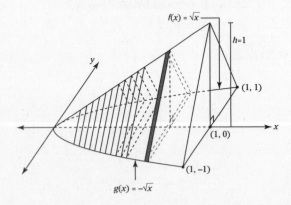

A. **The volume is ⅖ *cubic units*.**

1. **Always try to sketch the figure first (of course, I've done it for you here).**

2. **Indicate on your sketch a representative thin slice of the volume in question.**

   This slice should always be perpendicular to the axis or direction along which you are integrating. In other words, if your integrand contains, say, a *dx,* your slice should be perpendicular to the *x*-axis. Also, the slice should not be at either end of the 3-dimensional figure or at any other special place. Rather, it should be "in the middle of nowhere."

3. **Express the volume of this slice.**

   It's easy to show — trust me — that the height of each triangle is the same as its *x*-coordinate. Its base goes from $-\sqrt{x}$ up to $\sqrt{x}$ and is thus $2\sqrt{x}$. And its thickness is *dx*.

   Therefore,
   $Volume_{slice} = \frac{1}{2}\left(2\sqrt{x}\right)x \cdot dx = x\sqrt{x}\, dx$

4. **Add up the slices from 0 to 1 by integrating.**

   $$\int_0^1 x\sqrt{x}\, dx = \frac{2}{5}x^{5/2}\Big]_0^1 = \frac{2}{5}\ cubic\ units$$

EXAMPLE

Q. Using the disk/washer method, what's the volume of the glass that makes up the vase shown in the following figure?

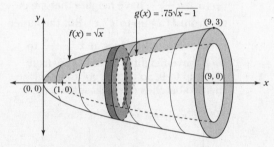

A. The volume is $\frac{45\pi}{2}$.

First, here's how the vase is "created." The light gray shaded area shown in the figure lies between $\sqrt{x}$ and $.75\sqrt{x-1}$ from $x = 0$ to $x = 9$. The three-dimensional vase shape is generated by revolving the shaded area about the x-axis.

1. **Sketch the 3-D shape (already done for you).**

2. **Indicate a representative slice (see the dark gray shaded area in the figure).**

3. **Express the volume of the representative slice.**

   A representative slice in a washer problem looks like — can you guess? — a washer. See the following figure.

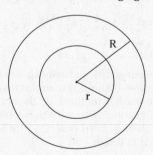

The large circle has an area of πR^2, and the hole an area of πr^2. So a washer's cross-sectional area is $\pi R^2 - \pi r^2$, or $\pi(R^2 - r^2)$. It's thickness is dx, so its volume is $\pi(R^2 - r^2)\,dx$.

Back to our problem. Big R in the vase problem is $\sqrt{x}$ and little r is $.75\sqrt{x-1}$, so the volume of a representative washer is $\pi\left(\left(\sqrt{x}\right)^2 - \left(.75\sqrt{x-1}\right)^2\right)dx$

4. **Add up the washers by integrating from 0 to 9.**

   But wait; did you notice the slight snag in this problem? The "washers" from $x = 0$ to $x = 1$ have no holes so there's no little-r circle to subtract from the big-R circle. A washer without a hole is called a disk, but you treat it the same as a washer except you don't subtract a hole.

5. **Add up the disks from 0 to 1 and the washers from 1 to 9 for the total volume.**

$$Volume_{vase} = \int_0^1 \pi\sqrt{x}^2\,dx + \int_1^9 \pi\left(\sqrt{x}^2 - \left(.75\sqrt{x-1}\right)^2\right)dx$$

$$= \pi\int_0^1 x\,dx + \pi\int_1^9 \left(x - \frac{9}{16}(x-1)\right)dx$$

$$= \pi\int_0^1 x\,dx + \pi\int_1^9 x\,dx - \frac{9\pi}{16}\int_1^9 (x-1)\,dx$$

$$= \pi\int_0^9 x\,dx - \frac{9\pi}{16}\int_1^9 (x-1)\,dx$$

$$= \frac{\pi}{2}x^2\Big|_0^9 - \frac{9\pi}{32}(x-1)^2\Big|_1^9$$

$$= \frac{81\pi}{2} - 18\pi$$

$$= \frac{45\pi}{2}$$

Q. Now tip the same glass vase up vertically. This time find the volume of its glass with the cylindrical shells method. See the following figure. (Did you notice that the shape of the vase is now somewhat different? Sorry about that.)

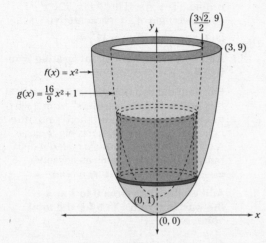

A. The volume is $\dfrac{45p}{2}$.

Again, this is the same vase as in the disk/washer example, but this time it's represented by different functions. In a random act of kindness, I figured the new functions for you.

1. **Express the volume of your representative shell.**

To figure the volume of a representative shell, imagine taking the label off a can of soup — it's a rectangle, right? The area is *base · height* and the base is the circumference of the can. So the area is $2\pi rh$. (r equals x and h depends on the given functions.) The thickness of the shell is dx, so its volume is $2\pi rh\,dx$.

Wait! Another snag — similar to but unrelated to the one in the last example. The smaller shells, with right edges at $x = 0$ up to $x = \dfrac{3\sqrt{2}}{2}$, have heights that measure from $f(x)$ up to $g(x)$. But the larger shells, with right edges at $x = \dfrac{3\sqrt{2}}{2}$ to $x = 3$, have heights that measure from $f(x)$ up to 9. So you've got to integrate the two batches of shells separately.

$Volume_{\text{smaller shells}} = 2\pi rh\,dx =$

$$2\pi x \left(\underbrace{\frac{16}{9}x^2 + 1}_{top\,(g(x))} - \underbrace{x^2}_{bottom\,(f(x))} \right) dx$$

$Volume_{\text{larger shells}} = 2\pi rh\,dx =$
$2\pi x \left(9 - x^2 \right) dx$

2. **Add up all the shells by integrating.**

With the cylindrical shells method, you integrate from the center to the outer edge.

$$\int_0^{3\sqrt{2}/2} 2\pi x \left(\frac{16}{9}x^2 - x^2 + 1 \right) dx + \int_{3\sqrt{2}/2}^3 2\pi x \left(9 - x^2 \right) dx$$

$$= 2\pi \int_0^{3\sqrt{2}/2} \left(\frac{7}{9}x^3 + x \right) dx + 2\pi \int_{3\sqrt{2}/2}^3 \left(-x^3 + 9x \right) dx$$

$$= 2\pi \left[\frac{7}{36}x^4 + \frac{1}{2}x^2 \right]_0^{3\sqrt{2}/2} + 2\pi \left[-\frac{1}{4}x^4 + \frac{9}{2}x^2 \right]_{3\sqrt{2}/2}^3$$

$$= 2\pi \left(\frac{63}{16} + \frac{9}{4} \right) + 2\pi \left(-\frac{81}{4} + \frac{81}{2} - \left(-\frac{81}{16} + \frac{81}{4} \right) \right)$$

$$= \frac{45\pi}{2}$$

Amazing! This actually agrees (which, of course, it should) with the result from the washer method. By the way, I got a bit carried away with these example problems. Your practice problems won't be this tough.

*7. Use the meat slicer method to derive the formula for the volume of a pyramid with a square base (see the following figure). Integrate from 0 to h along the positive side of the upside-down y-axis. (I set the problem up this way because it simplifies it. You can draw the y-axis the regular way if you like, but then you get an upside-down pyramid.)

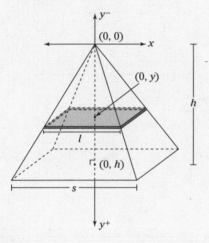

Solve It

8. Use the washer method to find the volume of the solid that results when the area enclosed by $f(x) = x$ and $g(x) = \sqrt{x}$ is revolved around the x-axis.

Solve It

9. Same as problem 8, but with $f(x) = x^2$ and $g(x) = 4x$.

Solve It

*10. Use the disk method to derive the formula for the volume of a cone. ***Hint:*** What's your function? See the following figure.

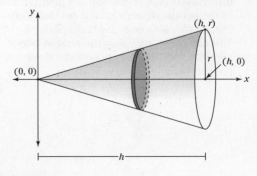

Solve It

11. Use the cylindrical shells method to find the volume of the solid that results when the area enclosed by $f(x) = x^2$ and $g(x) = x^3$ is revolved about the *y*-axis.

Solve It

*12. Use the cylindrical shells method to find the volume of the solid that results when the area enclosed by $\sin x$, $\cos x$, and the *x*-axis is revolved about the *y*-axis.

Solve It

Arc Length and Surfaces of Revolution

Integration determines the length of a curve by cutting it up into an infinite number of infinitesimal segments, each of which is sort of the hypotenuse of a tiny right triangle. Then your pedestrian Pythagorean Theorem does the rest. The same basic idea applies to surfaces of revolution. Here are two handy formulas for solving these problems:

✔ **Arc length:** The length along a function, $f(x)$, from a to b is given by

$$Arc\ Length = \int_a^b \sqrt{1 + \left(f'(x)\right)^2}\, dx$$

✔ **Surface of revolution:** The surface area generated by revolving the portion of a function, $f(x)$, between $x = a$ and $x = b$ about the x-axis is given by

$$Surface\ Area = 2\pi \int_a^b f(x)\sqrt{1 + \left(f'(x)\right)^2}\, dx$$

EXAMPLE

Q. What's the arc length along $f(x) = x^{2/3}$ from $x = 8$ to $x = 27$?

A. The arc length is about 19.65.

1. Find $f'(x)$.

$$f(x) = x^{2/3} \qquad f'(x) = \frac{2}{3} x^{-1/3}$$

2. Plug into the arc length formula.

$$Arc\ Length_{8\ to\ 27} = \int_8^{27} \sqrt{1 + \frac{4}{9} x^{-2/3}}\, dx$$

3. Integrate.

These arc length problems tend to produce tricky integrals; I'm not going to show all the work here.

$$= \frac{1}{3} \int_8^{27} \sqrt{9 + 4x^{-2/3}}\, dx$$

$$= \frac{1}{3} \int_8^{27} x^{-1/3} \sqrt{9x^{2/3} + 4}\, dx$$

You finish this with a u-substitution, where $u = 9x^{2/3} + 4$.

$$= \frac{1}{3} \int_{40}^{85} \frac{1}{6} u^{1/2}\, du$$

$$= \frac{1}{18} \left[\frac{2}{3} u^{3/2} \right]_{40}^{85}$$

$$= \frac{85\sqrt{85} - 80\sqrt{10}}{27}$$

$$\approx 19.65$$

An eminently sensible answer, because from $x = 8$ to $x = 27$, $x^{2/3}$ is very similar to a straight line of length $27 - 8$, which equals 19.

Q. Find the surface area generated by revolving $f(x) = \frac{1}{3} x^3$ ($0 \le x \le 2$) about the x-axis.

A. The area is $\frac{\pi}{9}\left(17\sqrt{17} - 1\right)$.

1. Find the function's derivative.

$$f(x) = \frac{1}{3} x^3 \qquad f'(x) = x^2$$

2. Plug into the surface area formula.

$$Surface\ Area = 2\pi \int_0^2 \frac{1}{3} x^3 \sqrt{1 + \left(x^2\right)^2}\, dx$$

$$= \frac{2\pi}{3} \int_0^2 x^3 \sqrt{1 + x^4}\, dx$$

You can do this integral with u-substitution.

$$u = 1 + x^4 \quad \text{when } x = 0,\ u = 1$$
$$du = 4x^3\, dx \quad \text{when } x = 2,\ u = 17$$

$$= \frac{2\pi}{3} \cdot \frac{1}{4} \int_0^2 4x^3 \sqrt{1 + x^4}\, dx$$

$$= \frac{\pi}{6} \int_1^{17} u^{1/2}\, du$$

$$= \frac{\pi}{6} \left[\frac{2}{3} u^{3/2} \right]_1^{17}$$

$$= \frac{\pi}{9}\left(17\sqrt{17} - 1\right)$$

13. Find the distance from (2, 1) to (5, 10) with the arc length formula.

Solve It

14. What's the surface area generated by revolving $f(x) = \frac{3}{4}x$ from $x = 0$ to $x = 4$ about the x-axis?

Solve It

15. a. Confirm your answer to problem 13 with the distance formula.

   b. Confirm your answer to problem 14 with the formula for the lateral area of a cone, $LA = \pi r\ell$, where ℓ is the slant height of the cone.

Solve It

16. What's the surface area generated by revolving $f(x) = \sqrt{x}$ from $x = 0$ to $x = 9$ about the x-axis?

Solve It

Getting Your Hopes Up with L'Hôpital's Rule

This powerful little rule enables you to easily compute limits that are either difficult or impossible without it.

L'Hôpital's Rule: When plugging the arrow-number into a limit expression gives you 0/0 or ∞/∞, you replace the numerator and denominator with their respective derivatives and do the limit problem again — repeating this process if necessary — until you arrive at a limit you can solve.

If you're wondering why this limit rule is in the middle of this chapter about integration, it's because you need L'Hôpital's Rule for the next section and the next chapter.

Q. What's $\lim\limits_{x \to \infty} \dfrac{x}{\log x}$?

A. The limit is ∞.

1. **Plug ∞ into x:**

   You get $\frac{\infty}{\infty}$. Not an answer, but just what you want for L'Hôpital's Rule.

2. **Replace the numerator and denominator of the limit fraction with their respective derivatives.**

   $$= \lim_{x \to \infty} \frac{1}{\dfrac{1}{x \ln 10}} = \lim_{x \to \infty} (x \ln 10)$$

3. **Now you can plug in.**

   $$= \infty \cdot \ln 10 = \infty$$

   Remember: If substituting the arrow-number into x gives you $\pm\infty \cdot 0$, $\infty - \infty$, $1^{\pm\infty}$, 0^0, or $\pm\infty^0$ — the so-called *unacceptable* forms — instead of one of the *acceptable* forms, $\dfrac{0}{0}$ or $\dfrac{\pm\infty}{\pm\infty}$, you have to manipulate the limit problem to convert it into one of the acceptable forms.

Q. What's $\lim\limits_{x \to \infty} (x^2 e^{-x})$?

A. The limit is 0.

1. **Plug ∞ into x.**

   You get $\infty \cdot 0$, one of the unacceptable forms.

2. **Rewrite e^{-x} as $\dfrac{1}{e^x}$ to produce an acceptable form: $\lim\limits_{x \to \infty} \dfrac{x^2}{e^x}$.**

   Plugging in now gives you what you need, $\frac{\infty}{\infty}$.

3. **Replace numerator and denominator with their derivatives.**

   $$= \lim_{x \to \infty} \frac{2x}{e^x}$$

4. **Plugging in gives you $\frac{\infty}{\infty}$ again, so you use L'Hôpital's Rule a second time.**

   $$= \lim_{x \to \infty} \frac{2}{e^x} = \frac{2}{e^\infty} = \frac{2}{\infty} = 0$$

17. What's $\lim\limits_{x \to \pi/2} \dfrac{\cos x}{x - \dfrac{\pi}{2}}$?

Solve It

18. $\lim\limits_{x \to 0} \dfrac{1 - \cos x}{x^2} = ?$

Solve It

19. Evaluate $\lim\limits_{x \to \pi/4} \big((\tan x - 1)\sec 6x \big)$.

Solve It

20. What's $\lim\limits_{x \to 0^+} \left(\dfrac{1}{x} + \dfrac{1}{\cos x - 1} \right)$?

Solve It

21. Evaluate $\lim\limits_{x \to 0^+} (\csc x - \log x)$.

Solve It

*22. What's $\lim\limits_{x \to 0} (1 + x)^{1/x}$? **Tip:** When plugging in gives you one of the exponential forms, $0^0, \infty^0$, or $1^{\pm\infty}$, set the limit equal to y, take the natural log of both sides, use the log of a power rule, and take it from there.

Solve It

Disciplining Those Improper Integrals

In this section, you bring some discipline to integrals that misbehave by going up, down, left, or right to infinity. You handle infinity, as usual, with limits. Here's an integral that goes up to infinity:

Q. Evaluate $\int_{-1}^{2} \frac{1}{x^2}\, dx$.

A. **The area is infinite.**

1. **Check whether the function is defined everywhere between and at the limits of integration.**

   You note that when $x = 0$, the function shoots up to infinity. So you've got an improper integral. In a minute, you'll see what happens if you fail to note this.

2. **Break the integral in two at the critical x-value.**

   $$\int_{-1}^{2} \frac{1}{x^2}\, dx = \int_{-1}^{0} \frac{1}{x^2}\, dx + \int_{0}^{2} \frac{1}{x^2}\, dx$$

3. **Replace the critical x-value with constants and turn each integral into a limit.**

   $$= \lim_{a \to 0^-} \int_{-1}^{a} \frac{1}{x^2}\, dx + \lim_{b \to 0^+} \int_{b}^{2} \frac{1}{x^2}\, dx$$

4. **Integrate.**

   $$= \lim_{a \to 0^-} \left[-\frac{1}{x} \right]_{-1}^{a} + \lim_{b \to 0^+} \left[-\frac{1}{x} \right]_{b}^{2}$$

   $$= \lim_{a \to 0^-} \left(-\frac{1}{a} - \left(-\frac{1}{-1} \right) \right) + \lim_{b \to 0^+} \left(-\frac{1}{2} - \left(-\frac{1}{b} \right) \right)$$

   $$= \infty + \infty = \infty$$

   Therefore, this limit does not exist (DNE).

   Warning: If you split up an integral in two and one piece equals ∞ and the other equals $-\infty$, you *cannot* add the two to obtain an answer of zero. When this happens, the limit DNE.

   Now, watch what happens if you fail to notice that this function is undefined at $x = 0$.

   $$\int_{-1}^{2} \frac{1}{x^2}\, dx = -\frac{1}{x} \Big]_{-1}^{2} = -\frac{1}{2} - \left(-\frac{1}{-1} \right) = -\frac{3}{2}$$

   Wrong! (And absurd, because the function is positive everywhere from $x = -1$ to $x = 2$.)

Q. Evaluate $\int_1^\infty \frac{1}{x^2}\,dx$.

A. The area is 1.

1. Replace ∞ with c, and turn the integral into a limit.

$$\lim_{c \to \infty} \int_1^c \frac{1}{x^2}\,dx$$

2. Integrate.

$$= \lim_{c \to \infty}\left[-\frac{1}{x}\right]_1^c = \lim_{c \to \infty}\left(-\frac{1}{c} - (-1)\right) = -\frac{1}{\infty} + 1 = 1$$

Amazing! This infinitely long sliver of area has an area of 1 square unit.

23. Evaluate $\int_{-32}^1 \frac{dx}{\sqrt[5]{x}}$.

Solve It

24. Compute $\int_0^6 x \ln x\,dx$.

Solve It

25. $\int\limits_{1}^{\infty} \dfrac{dx}{x\sqrt{x^2-1}} =$? **Hint:** Split up at $x = 2$.

Solve It

26. What's $\int\limits_{1}^{\infty} \dfrac{1}{x}\, dx$?

Solve It

*27. $\int\limits_{1}^{\infty} \dfrac{1}{x}\sqrt{\arctan x}\, dx =$? **Hint:** Use problem 26.

Solve It

*28. $\int\limits_{-\infty}^{\infty} \dfrac{1}{x}\, dx =$? **Hint:** Break into four parts.

Solve It

Solutions to Integration Application Problems

1   What's the average value of $f(x) = \dfrac{x}{(x^2+1)^3}$ from 1 to 3? **The average value is 0.03.**

$$Ave\ value = \frac{total\ area}{base} = \frac{\displaystyle\int_1^3 \frac{x}{(x^2+1)^3}}{3-1}$$

TIP

Do this with a u-substitution.

$$= \frac{\dfrac{1}{2}\displaystyle\int_1^3 \dfrac{2x}{(x^2+1)^3}\,dx}{2}$$

$u = x^2+1$ when $x=1$, $u=2$ $= \dfrac{1}{4}\displaystyle\int_2^{10} \dfrac{du}{u^3}$

$du = 2x\,dx$ when $x=3$, $u=10$

$$= -\frac{1}{8}\left[u^{-2}\right]_2^{10}$$

$$= -\frac{1}{8}\left(10^{-2} - 2^{-2}\right)$$

$$= 0.03$$

2   A car's speed in feet per second is given by $f(t) = t^{1.7} - 6t + 80$. What's its average speed from $t = 5$ seconds to $t = 15$ seconds? What's that in miles per hour? $\approx$ **49.51 *miles per hour*.**

$$Ave\ speed = \frac{total\ distance}{total\ time} = \frac{\displaystyle\int_5^{15}\left(t^{1.7} - 6t + 80\right)dt}{15-5}$$

$$= \frac{\dfrac{1}{2.7}t^{2.7} - 3t^2 + 80t \Big]_5^{15}}{10}$$

$$\approx \frac{554.73 - 675 + 1200 - (28.57 - 75 + 400)}{10}$$

$$\approx 72.616\ feet\ per\ second$$

$$\approx 49.51\ miles\ per\ hour$$

3   What's the area enclosed by $f(x) = x^2$ and $g(x) = \sqrt{x}$? **The area is ⅓.**

1. Graph the functions.

2. Find the points of intersection. I presume you had no trouble finding them: (0, 0) and (1, 1).

3. Find the area. ***Remember:*** top minus bottom.

$$Area = \int_0^1 \left(\sqrt{x} - x^2\right)dx = \frac{2}{3}x^{3/2} - \frac{1}{3}x^3 \Big]_0^1 = \frac{2}{3} - \frac{1}{3} = \frac{1}{3}$$

4   What's the total area enclosed by $f(t) = t^3$ and $g(t) = t^5$? **The area is ⅙.**

1. Graph the functions. You should see three points of intersection.

2. Find the points: (–1, –1), (0, 0), and (1, 1).

3. Find the area on the left.

t^5 is on top, so

$$Area = \int_{-1}^0 \left(t^5 - t^3\right)dt = \frac{1}{6}t^6 - \frac{1}{4}t^4 \Big]_{-1}^0 = 0 - \left(\frac{1}{6} - \frac{1}{4}\right) = \frac{1}{12}$$

4. Find the area on the right — t^3 is on top — then add that to $\frac{1}{12}$.

$$Area = \int_0^1 (t^3 - t^5) = \frac{1}{4} t^4 - \frac{1}{6} t^6 \Big|_0^1 = \frac{1}{4} - \frac{1}{6} = \frac{1}{12}$$

Therefore, the total area is $\frac{1}{12} + \frac{1}{12}$, or $\frac{1}{6}$.

Note that had you observed that both t^3 and t^5 are odd functions, you could have reasoned that the two areas are the same, and then calculated just one of them and doubled the result.

*5 The lines $y = x$, $y = 2x - 5$, and $y = -2x + 3$ form a triangle in the first and fourth quadrants. What's the area of this triangle? **The area is 6.**

1. Graph the three lines.

2. Find the three points of intersection.

a. $y = x$ intersects $\qquad$ **b.** $y = x$ intersects $\qquad$ **c.** $y = 2x - 5$ intersects

$y = 2x - 5$ at $x = 2x - 5$ $\qquad$ $y = -2x + 3$ at $x = -2x + 3$ $\qquad$ $y = -2x + 3$ at $2x - 5 = -2x + 3$

$x = 5$ and, thus, $y = 5$ $\qquad$ $x = 1$ and, thus, $y = 1$ $\qquad$ $x = 2$ and, thus, $y = -1$

3. Integrate to find the area from $x = 1$ to $x = 2$; $y = x$ is on the top and $y = -2x + 3$ is on the bottom, so

$$Area = \int_1^2 \left(x - (-2x + 3) \right) dx$$

$$= 3 \int_1^2 (x - 1) \, dx$$

$$= 3 \left[\frac{1}{2} x^2 - x \right]_1^2$$

$$= 3 \left[(2 - 2) - \left(\frac{1}{2} - 1 \right) \right] = \frac{3}{2}$$

4. Integrate to find the area from $x = 2$ to $x = 5$; $y = x$ is on the top again, but $y = 2x - 5$ is on the bottom, thus

$$Area = \int_2^5 \left(x - (2x - 5) \right) dx$$

$$= \int_2^5 (-x + 5) \, dx$$

$$= -\frac{1}{2} x^2 + 5x \Big|_2^5$$

$$= -\frac{25}{2} + 25 - (-2 + 10) = \frac{9}{2}$$

Grand total from Steps 3 and 4 equals 6.

Granted, using calculus for this problem is loads of fun, but it's totally unnecessary. If you cut the triangle into two triangles — corresponding to Steps 3 and 4 above — you can get the total area with simple coordinate geometry.

6 What's the area of the triangular shape in the first quadrant enclosed by $\sin x$, $\cos x$, and the line $y = \frac{1}{2}$? **The area is** $\sqrt{3} - \sqrt{2} - \frac{\pi}{12}$.

1. Do the graph and find the intersections.

   a. From the example, you know that $\sin x$ and $\cos x$ intersect at $x = \frac{\pi}{4}$.

   b. $y = \frac{1}{2}$ intersects $\sin x$ at $\sin x = \frac{1}{2}$ so $x = \frac{\pi}{6}$.

   c. $y = \frac{1}{2}$ intersects $\cos x$ at $\cos x = \frac{1}{2}$ so $x = \frac{\pi}{3}$.

2. Integrate to find the area between $\frac{\pi}{6}$ to $\frac{\pi}{4}$ and between $\frac{\pi}{4}$ to $\frac{\pi}{3}$.

$$Area = \int\limits_{\pi/6}^{\pi/4} \left(\sin x - \frac{1}{2} \right) dx + \int\limits_{\pi/4}^{\pi/3} \left(\cos x - \frac{1}{2} \right) dx$$

$$= -\cos x - \frac{1}{2} x \Big]_{\pi/6}^{\pi/4} + \sin x - \frac{1}{2} x \Big]_{\pi/4}^{\pi/3}$$

$$= -\frac{\sqrt{2}}{2} - \frac{\pi}{8} - \left(-\frac{\sqrt{3}}{2} - \frac{\pi}{12} \right) + \frac{\sqrt{3}}{2} - \frac{\pi}{6} - \left(\frac{\sqrt{2}}{2} - \frac{\pi}{8} \right)$$

$$= \sqrt{3} - \sqrt{2} - \frac{\pi}{12} \quad \text{Cool answer, eh?}$$

*7 Use the meat slicer method to derive the formula for the volume of a pyramid with a square base. **The formula is** $\frac{1}{3} s^2 h$.

Using similar triangles, you can establish the following proportion: $\frac{y}{h} = \frac{l}{s}$.

You want to express the side of your representative slice as a function of y (and the constants, s and h), so that's $l = \frac{ys}{h}$.

The volume of your representative square slice equals its cross-sectional area times its thickness, dy, so now you've got

$$Volume_{slice} = \left(\frac{ys}{h} \right)^2 dy$$

Don't forget that when integrating, constants behave just like ordinary numbers.

$$Volume_{pyramid} = \int\limits_{0}^{h} \left(\frac{ys}{h} \right)^2 dy = \frac{s^2}{h^2} \int\limits_{0}^{h} y^2 \, dy = \frac{s^2}{h^2} \cdot \frac{1}{3} y^3 \Big]_{0}^{h} = \frac{s^2}{h^2} \cdot \frac{1}{3} h^3 = \frac{1}{3} s^2 h$$

That's the old familiar pyramid formula: $\frac{1}{3} \cdot base \cdot height$ — the hard way.

8 Use the washer method to find the volume of the solid that results when the area enclosed by $f(x) = x$ and $g(x) = \sqrt{x}$ is revolved about the x-axis. **The volume is** $\frac{\pi}{6}$.

1. Sketch the solid, including a representative slice. See the following figure.

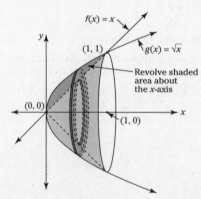

2. Express the volume of your representative slice.

$$Volume_{washer} = \pi \left(R^2 - r^2 \right) dx = \pi \left(\sqrt{x}^2 - x^2 \right) dx = \pi \left(x - x^2 \right) dx$$

3. Add up the infinite number of infinitely thin washers from 0 to 1 by integrating.

$$Volume_{solid} = \int_0^1 \pi \left(x - x^2 \right) dx = \pi \left[\frac{1}{2} x^2 - \frac{1}{3} x^3 \right]_0^1 = \pi \left(\frac{1}{2} - \frac{1}{3} \right) = \frac{\pi}{6}$$

Note that the infinite number of washers you just added contain an infinite number of holes — *way* more than the number of holes it takes to fill the Albert Hall. For extra credit: What "holes" were the Beatles referring to? ***Hint:*** Remember the charioteer Glutius from Chapter 8?

9 Same as problem 8, but with $f(x) = x^2$ and $g(x) = 4x$. **The volume is $\frac{2048\pi}{15}$ cubic units.**

1. Sketch the solid and a representative slice. See the following figure.

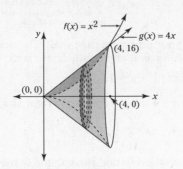

2. Determine where the functions intersect.

$$x^2 = 4x$$
$$x = 4 \text{ and thus } y = 16$$

3. Express the volume of a representative washer.

$$Volume_{washer} = \pi \left(R^2 - r^2 \right) dx = \pi \left(\left(4x \right)^2 - \left(x^2 \right)^2 \right) dx = \pi \left(16x^2 - x^4 \right) dx$$

4. Add up the washers from 0 to 4 by integrating.

$$Volume_{solid} = \pi \int_0^4 \left(16x^2 - x^4 \right) dx = \pi \left[\frac{16}{3} x^3 - \frac{1}{5} x^5 \right]_0^4 = \pi \left(\frac{1024}{3} - \frac{1024}{5} \right) = \frac{2048\pi}{15}$$

*10 Use the disk method to derive the formula for the volume of a cone. **The formula is $\frac{1}{3}\pi r^2 h$.**

1. Find the function that revolves about the x-axis to generate the cone.

The function is the line that goes through $(0, 0)$ and (h, r). Its slope is $\frac{r}{h}$ and thus its equation is $f(x) = \frac{r}{h} x$.

2. Express the volume of a representative disk. The radius of your representative disk is $f(x)$ and its thickness is dx. Its volume is therefore

$$V_{disk} = \pi \left(f(x) \right)^2 dx = \pi \left(\frac{r}{h} x \right)^2 dx$$

3. Add up the disks from $x = 0$ to $x = h$ by integrating. Don't forget that r and h are simple constants.

$$V_{cone} = \int_0^h \pi \left(\frac{r}{h} x \right)^2 dx = \frac{\pi r^2}{h^2} \int_0^h x^2 dx = \frac{\pi r^2}{h^2} \left[\frac{1}{3} x^3 \right]_0^h = \frac{\pi r^2}{h^2} \cdot \frac{1}{3} h^3 = \frac{1}{3} \pi r^2 h$$

11 Use the cylindrical shells method to find the volume of the solid that results when the area enclosed by $f(x) = x^2$ and $g(x) = x^3$ is revolved about the y-axis. **The volume is $\frac{\pi}{10}$.**

1. Sketch your solid. See the following figure.

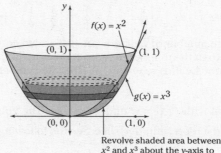

Revolve shaded area between
x^2 and x^3 about the y-axis to
create a bowl-like shape.

2. Express the volume of your representative shell. The height of the shell equals *top* minus *bottom*, or $x^2 - x^3$. Its radius is x, and its thickness is dx. Its volume is thus

$$Volume_{shell} = 2\pi rh\,dx = 2\pi x\left(x^2 - x^3\right)dx$$

3. Add up the shells from $x = 0$ to $x = 1$ (center to right end) by integrating.

$$Volume_{bowl} = 2\pi \int_0^1 \left(x^3 - x^4\right)dx = 2\pi\left[\frac{1}{4}x^4 - \frac{1}{5}x^5\right]_0^1 = \frac{\pi}{10}$$

*12 Use the cylindrical shells method to find the volume of the solid that results when the area enclosed by $\sin x$, $\cos x$, and the x-axis is revolved about the y-axis. **The volume is $\pi^2 - \frac{\pi^2\sqrt{2}}{2}$.**

1. Sketch the dog bowl. See the following figure.

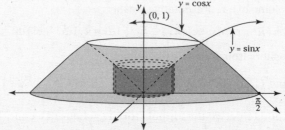

2. Determine where the two functions cross. You should obtain $\left(\frac{\pi}{4}, \frac{\sqrt{2}}{2}\right)$.

3. Express the volume of your representative shell. I'm sure you noticed that the shells with a radius less than $\frac{\pi}{4}$ have a height of $\sin x$, while the larger shells have a height of $\cos x$. So you have to add up two batches of shells:

$$Volume_{smaller\ shell} = 2\pi rh\,dx$$
$$= 2\pi x \sin x\,dx$$
$$Volume_{larger\ shell} = 2\pi x \cos x\,dx$$

4. Add up the two batches of shells.

$$Volume_{dog\ bowl} = 2\pi \int_0^{\pi/4} x \sin x\,dx + 2\pi \int_{\pi/4}^{\pi/2} x \cos x\,dx$$

Both of these integrals are easy to do with the integration by parts method with $u = x$ in both cases. I'll leave it up to you. You should obtain the following:

$$= 2\pi \left[-x\cos x + \sin x \right]_0^{\pi/4} + 2\pi \left[x\sin x + \cos x \right]_{\pi/4}^{\pi/2}$$

$$= 2\pi \left(-\frac{\pi}{4} \cdot \frac{\sqrt{2}}{2} + \frac{\sqrt{2}}{2} \right) + 2\pi \left(\frac{\pi}{2} - \frac{\pi}{4} \cdot \frac{\sqrt{2}}{2} - \frac{\sqrt{2}}{2} \right)$$

$$= \pi^2 - \frac{\pi^2 \sqrt{2}}{2}$$

13 Find the distance from (2, 1) to (5, 10) with the arc length formula. **The distance is $3\sqrt{10}$.**

1. Find a function for the "arc" — it's really a line, of course — that connects the two points. I'm sure you remember the point slope formula from your algebra days:

$$y - y_1 = m(x - x_1)$$
$$y - 1 = 3(x - 2)$$
$$y = 3x - 5$$

2. "Find" y' — I hope you don't have to look very far: $y' = 3$.

3. Plug into the formula: $Arc\ Length = \int_2^5 \sqrt{1 + 3^2}\, dx = x\sqrt{10}\Big]_2^5 = 3\sqrt{10}$.

14 What's the surface area generated by revolving $f(x) = \frac{3}{4}x$ from $x = 0$ to $x = 4$ about the x-axis? The surface area is **15π**.

1. Sketch the function and the surface.

2. Plug the function and its derivative into the formula.

$$SA = 2\pi \int_0^4 \frac{3}{4}x \sqrt{1 + \left(\frac{3}{4}\right)^2}\, dx = \frac{3\pi}{2} \int_0^4 x \sqrt{\frac{25}{16}}\, dx = \frac{15\pi}{8} \left[\frac{1}{2}x^2 \right]_0^4 = 15\pi$$

15 a. Confirm your answer to problem 13 with the distance formula.

$$d = \sqrt{(x_2 - x_1)^2 + (y_2 - y_1)^2} = \sqrt{(5 - 2)^2 + (10 - 1)^2} = \mathbf{3\sqrt{10}}$$

b. Confirm your answer to problem 14 with the formula for the lateral area of a cone, $LA = \pi r \ell$, where ℓ is the slant height of the cone.

1. Determine the radius and slant height of the cone.

From your sketch and the function, you can easily determine that the function goes through (4, 3), and that, therefore, the radius is 3 and the slant height is 5 (it's the hypotenuse of a 3-4-5 triangle).

2. Plug into the formula.

Lateral Area $= \pi r \ell = \mathbf{15\pi}$ It checks.

16 What's the surface area generated by revolving $f(x) = \sqrt{x}$ from $x = 0$ to $x = 9$ about the x-axis? **The surface area is $\frac{\pi}{6}\left(37\sqrt{37} - 1\right)$.**

1. Plug the formula and its derivative into the formula.

$$f(x) = \sqrt{x} \qquad f'(x) = \frac{1}{2\sqrt{x}}$$

$$Surface\ Area = 2\pi \int_0^9 \sqrt{x} \sqrt{1 + \left(\frac{1}{2\sqrt{x}}\right)^2}\, dx = 2\pi \int_0^9 \sqrt{x} \sqrt{1 + \frac{1}{4x}}\, dx = 2\pi \int_0^9 \sqrt{x + \frac{1}{4}}\, dx$$

2. Integrate. $= 2\pi \left[\frac{2}{3}\left(x + \frac{1}{4}\right)^{3/2} \right]_0^9 = \frac{4\pi}{3}\left(\left(\frac{37}{4}\right)^{3/2} - \left(\frac{1}{4}\right)^{3/2}\right) = \frac{\pi}{6}\left(37\sqrt{37} - 1\right)$

17 $\lim\limits_{x \to \pi/2} \dfrac{\cos x}{x - \dfrac{\pi}{2}} = -1$

1. Plug in: $\dfrac{0}{0}$ — onward!

2. Replace numerator and denominator with their derivatives: $= \lim\limits_{x \to \pi/2} \dfrac{-\sin x}{1}$.

3. Plug in again: $= \dfrac{-\sin \dfrac{\pi}{2}}{1} = -1$.

18 $\lim\limits_{x \to 0} \dfrac{1 - \cos x}{x^2} = \dfrac{1}{2}$

1. Plug in: $\dfrac{0}{0}$; no worries.

2. Replace with derivatives: $= \lim\limits_{x \to 0} \dfrac{\sin x}{2x}$.

3. Plug in: $\dfrac{0}{0}$ again, so repeat.

4. Replace with derivatives again: $= \lim\limits_{x \to 0} \dfrac{\cos x}{2}$.

5. Finish: $= \dfrac{1}{2}$.

19 $\lim\limits_{x \to \pi/4} \big((\tan x - 1) \sec 6x \big) = \dfrac{1}{3}$

1. Plugging in gives you $0 \cdot \infty$, so on to Step 2.

2. Rewrite: $= \lim\limits_{x \to \pi/4} \dfrac{\tan x - 1}{\cos 6x} = \dfrac{0}{0}$: copasetic.

3. Replace with derivatives: $= \lim\limits_{x \to \pi/4} \dfrac{\sec^2 x}{-6 \sin 6x}$.

4. Plug in to finish: $= \dfrac{\sec^2 \dfrac{\pi}{4}}{-6 \sin \dfrac{3\pi}{2}} = \dfrac{2}{6} = \dfrac{1}{3}$.

20 $\lim\limits_{x \to 0^+} \left(\dfrac{1}{x} + \dfrac{1}{\cos x - 1} \right) = -\infty$

1. Plugging in gives you $\infty - \infty$; no good.

2. Rewrite by adding the fractions: $= \lim\limits_{x \to 0^+} \dfrac{\cos x - 1 + x}{x(\cos x - 1)}$. That's a good bingo: $\dfrac{0}{0}$.

3. Replace with derivatives: $= \lim\limits_{x \to 0^+} \dfrac{-\sin x + 1}{(\cos x - 1) - x \sin x}$.

4. Plug in to finish: $= \dfrac{1}{"-0"} = -\infty$.

This 0 is "negative" because the denominator in the line just above is negative when x is approaching zero from the right. By the way, don't use "–0" in class — your teacher will call a technical on you.

21 $\lim\limits_{x \to 0^+} \big(\csc x - \log x \big) = \infty$

1. This limit equals $\infty - (-\infty)$, which equals $\infty + \infty = \infty$.

2. You're done! L'Hôpital's Rule isn't needed. You gotta be on your toes.

*22 $\lim\limits_{x \to 0} (1 + x)^{1/x} = e$

1. This is a 1^∞ case — time for a new technique.

2. Set your limit equal to y and take the natural log of both sides.

$$y = \lim\limits_{x \to 0} (1 + x)^{1/x}$$
$$\ln y = \ln \left(\lim\limits_{x \to 0} (1 + x)^{1/x} \right)$$

3. I give you permission to pull the limit to the outside: $\ln y = \lim\limits_{x \to 0}\left(\ln\left(1+x\right)^{1/x}\right)$.

4. Use the log of a power rule: $\ln y = \lim\limits_{x \to 0}\left(\frac{1}{x}\ln\left(1+x\right)\right)$.

5. You've got a $\infty \cdot 0$ case so rewrite: $\ln y = \lim\limits_{x \to 0}\dfrac{\ln\left(1+x\right)}{x}$.

6. You get $\frac{0}{0}$ — I'm down with it.

7. Replace with derivatives: $\ln y = \lim\limits_{x \to 0}\dfrac{\frac{1}{1+x}}{1} = 1$.

8. Your original limit equals y, so you've got to solve for y.

$$\ln y = 1$$
$$y = e$$

23 Evaluate $\displaystyle\int_{-32}^{1}\frac{dx}{\sqrt[5]{x}} = -\mathbf{18.75}$.

1. Undefined at $x = 0$, so break in two: $\displaystyle\int_{-32}^{1}\frac{dx}{\sqrt[5]{x}} = \int_{-32}^{0}\frac{dx}{\sqrt[5]{x}} + \int_{0}^{1}\frac{dx}{\sqrt[5]{x}}$.

2. Turn into limits: $= \lim\limits_{a \to 0^-}\displaystyle\int_{-32}^{a}\frac{dx}{\sqrt[5]{x}} + \lim\limits_{b \to 0^+}\int_{b}^{1}\frac{dx}{\sqrt[5]{x}}$.

3. Integrate: $= \lim\limits_{a \to 0^-}\left[\frac{5}{4}x^{4/5}\right]_{-32}^{a} + \lim\limits_{b \to 0^+}\left[\frac{5}{4}x^{4/5}\right]_{b}^{1} = 0 - \frac{5}{4}\cdot 16 + \frac{5}{4} - 0 = -18.75$.

24 $\displaystyle\int_{0}^{6} x\ln x\, dx = \mathbf{18\ln 6 - 9}$

1. The integral is improper because it's undefined at $x = 0$, so turn it into a limit:

$$= \lim\limits_{c \to 0^+}\int_{c}^{6} x\ln x\, dx$$

2. Integrate by parts. **Hint:** $\ln x$ is L from LIATE. You should obtain:

$$= \lim\limits_{c \to 0^+}\left[\frac{1}{2}x^2\ln x - \frac{1}{4}x^2\right]_{c}^{6}$$
$$= \lim\limits_{c \to 0^+}\left(\frac{1}{2}\cdot 36\ln 6 - 9 - \frac{1}{2}c^2\ln c + \frac{1}{4}c^2\right)$$
$$= 18\ln 6 - 9 - \frac{1}{2}\lim\limits_{c \to 0^+}\left(c^2\ln c\right)$$

3. Time to practice L'Hôpital's Rule. This is a $0 \cdot (-\infty)$ limit, so turn it into a $\frac{-\infty}{\infty}$ one:

$$= 18\ln 6 - 9 - \frac{1}{2}\lim\limits_{c \to 0^+}\dfrac{\ln c}{\dfrac{1}{c^2}}$$

4. Replace numerator and denominator with derivatives and finish:

$$= 18\ln 6 - 9 - \frac{1}{2}\lim\limits_{c \to 0^+}\dfrac{\dfrac{1}{c}}{-\dfrac{2}{c^3}} = 18\ln 6 - 9 - \frac{1}{2}\lim\limits_{c \to 0^+}\left(-\dfrac{c^2}{2}\right) = 18\ln 6 - 9$$

25 $\displaystyle\int_{1}^{\infty}\frac{dx}{x\sqrt{x^2-1}} = \frac{\pi}{2}$

This is a doubly improper integral because it goes up *and* right to infinity. You've got to split it up and tackle each infinite impropriety separately.

1. It doesn't matter where you split it up; how about 2, a nice, easy-to-deal-with number.

$$= \int_{1}^{2}\frac{dx}{x\sqrt{x^2-1}} + \int_{2}^{\infty}\frac{dx}{x\sqrt{x^2-1}}$$

2. Turn into limits.

$$= \lim_{a \to 1^+} \int_a^2 \frac{dx}{x\sqrt{x^2 - 1}} + \lim_{b \to \infty} \int_2^b \frac{dx}{x\sqrt{x^2 - 1}}$$

3. Integrate.

$$= \lim_{a \to 1^+} \left[\operatorname{arcsec} x\right]_a^2 + \lim_{b \to \infty} \left[\operatorname{arcsec} x\right]_2^b$$

$$= \lim_{a \to 1^+} \left(\operatorname{arcsec} 2 - \operatorname{arcsec} a\right) + \lim_{b \to \infty} \left(\operatorname{arcsec} b - \operatorname{arcsec} 2\right)$$

$$= \operatorname{arcsec} 2 - 0 + \frac{\pi}{2} - \operatorname{arcsec} 2$$

$$= \frac{\pi}{2}$$

26 $\int_1^\infty \frac{1}{x} \, dx = \infty$

1. Turn into a limit: $\lim_{c \to \infty} \int_1^c \frac{1}{x} \, dx$.

2. Integrate and finish: $= \lim_{c \to \infty} \left[\ln x\right]_1^c = \lim_{c \to \infty} (\ln c - \ln 1) = \infty$.

*27 $\int_1^\infty \frac{1}{x} \sqrt{\operatorname{arctan} x} \, dx = \infty$

No work is required for this one, "just" logic. You know from problem 26 that $\int_1^\infty \frac{1}{x} \, dx = \infty$.

Now, compare $\int_1^\infty \frac{1}{x} \sqrt{\operatorname{arctan} x} \, dx$ to $\int_1^\infty \frac{1}{x} \, dx$. But first note that because $\int_1^\infty \frac{1}{x} \, dx$ equals infinity, so will $\int_{10}^\infty \frac{1}{x} \, dx$, $\int_{100}^\infty \frac{1}{x} \, dx$, or $\int_{1,000,000}^\infty \frac{1}{x} \, dx$, because the area under $\frac{1}{x}$ from 1 to any other number must be finite.

From $\sqrt{3}$ to ∞, $\operatorname{arctan} x \geq 1$, and therefore $\sqrt{\operatorname{arctan} x} \geq 1$, and thus $\frac{1}{x}\sqrt{\operatorname{arctan} x} \geq \frac{1}{x}$. Finally, because $\int_{\sqrt{3}}^\infty \frac{1}{x} \, dx = \infty$ and because between $\sqrt{3}$ and ∞, $\frac{1}{x}\sqrt{\operatorname{arctan} x}$ is always equal to or greater than $\frac{1}{x}$, $\int_{\sqrt{3}}^\infty \frac{1}{x}\sqrt{\operatorname{arctan} x} \, dx$ must also equal ∞ and so, therefore, does $\int_1^\infty \frac{1}{x}\sqrt{\operatorname{arctan} x} \, dx$. Aren't you glad no work was required for this problem?

*28 $\int_{-\infty}^\infty \frac{1}{x} \, dx$ **is undefined.**

Quadruply improper!

1. Split into four parts: $\int_{-\infty}^\infty \frac{1}{x} \, dx = \int_{-\infty}^{-1} \frac{1}{x} \, dx + \int_{-1}^0 \frac{1}{x} \, dx + \int_0^1 \frac{1}{x} \, dx + \int_1^\infty \frac{1}{x} \, dx$.

2. Turn into limits: $= \lim_{a \to -\infty} \int_a^{-1} \frac{dx}{x} + \lim_{b \to 0^-} \int_{-1}^b \frac{dx}{x} + \lim_{c \to 0^+} \int_c^1 \frac{dx}{x} + \lim_{d \to \infty} \int_1^d \frac{dx}{x}$.

3. Integrate:

$$= \lim_{a \to -\infty} \left[\ln|x|\right]_a^{-1} + \lim_{b \to 0^-} \left[\ln|x|\right]_{-1}^b + \lim_{c \to 0^+} \left[\ln|x|\right]_c^1 + \lim_{d \to \infty} \left[\ln|x|\right]_1^d$$

$$= \lim_{a \to \infty} \left(\ln 1 - \ln|a|\right) + \lim_{b \to 0^-} \left(\ln|b| - \ln 1\right) + \lim_{c \to 0^+} \left(\ln 1 - \ln|c|\right) + \lim_{d \to \infty} \left(\ln|d| - \ln 1\right)$$

4. Finish: $= -\infty + (-\infty) + \infty + \infty$. Therefore, the limit doesn't exist, and the definite integral is thus undefined.

If you look at the graph of $y = \frac{1}{x}$, its perfect symmetry may make you think that $\int_{-\infty}^\infty \frac{1}{x} \, dx$ would equal zero. But — strange as it seems — it doesn't work that way.

Chapter 13

Infinite Series: Welcome to the Outer Limits

· ·

In This Chapter

▶ Twilight zone stuff

▶ Serious series

▶ Tests, tests, and more tests

· ·

*I*n this chapter, you look at something that's really quite amazing if you stop to think about it: sums of numbers that never end. Seriously, the sums of numbers in this chapter — if written out completely — would not fit in our universe. But despite the never-ending nature of these sums, some of them add up to a finite number! These are called *convergent* series. The rest are called *divergent*. Your task in this chapter is to decide which are which.

The Nifty nth Term Test

Because the mere beginning terms of any given sequence would completely fill the universe, and because the *n*th term is way beyond that, where is it? Does it really exist or is it only a figment of your imagination? If a tree falls in a forest and no one's there to hear it, does it make a sound?

First, a couple definitions. A *sequence* is a finite or infinite list of numbers (we will be dealing only with infinite sequences). When you add up the terms of a sequence, the sequence becomes a *series*. For example,

1, 2, 4, 8, 16, 32, 64, . . . is a sequence, and

1 + 2 + 4 + 8 + 16 + 32 + 64 + . . . is the related series.

If $\lim_{n \to \infty} a_n \neq 0$, then $\sum a_n$ diverges. In English, this says that if a series' underlying sequence does not converge to zero, then the series must diverge.

It does *not* follow that if a series' underlying sequence converges to zero then the series will definitely converge. It may converge, but there's no guarantee.

Q. Does $\sum_{n=1}^{\infty} \sqrt[n]{1 - \frac{1}{n}}$ converge or diverge?

A. **It diverges.**

You can answer this question with common sense if your calc teacher allows such a thing. As n gets larger and larger, $1 - \frac{1}{n}$ increases and gets closer and closer to one. And when you take any root of a number like 0.9, it gets *bigger* — and the higher the root index, the bigger the answer is. So $\sqrt[n]{1 - \frac{1}{n}}$ has to get larger as n increases and thus $\lim_{n \to \infty} \sqrt[n]{1 - \frac{1}{n}}$ cannot possibly equal zero. The series, therefore, diverges by the nth term test.

If your teacher is a stickler for rigor, you can do the following: Plugging in ∞ produces $\left(1 - \frac{1}{\infty}\right)^{1/\infty}$, which is 1^0, and that equals 1 — you're done. (Note that 1^0 is *not* one of the forms that gives you a L'Hôpital's Rule problem — see Chapter 12.) Because $\lim_{n \to \infty} \sqrt[n]{1 - \frac{1}{n}} = 1$, $\sum_{n=1}^{\infty} \sqrt[n]{1 - \frac{1}{n}}$ diverges.

1. Does $\sum_{n=1}^{\infty} \frac{2n^2 - 9n - 8}{5n^2 + 20n + 12}$ converge or diverge?

Solve It

2. Does $\sum_{n=1}^{\infty} \frac{1}{n}$ converge or diverge?

Solve It

Testing Three Basic Series

In this section, you figure out whether geometric series, *p*-series, and telescoping series are convergent or divergent.

- **Geometric series:** If $0 < |r| < 1$, the geometric series $\sum_{n=0}^{\infty} ar^n$ converges to $\frac{a}{1-r}$. If $|r| \geq 1$, the series diverges. Have you heard the riddle about walking halfway to the wall, then halfway again, then half the remaining distance, and so on? Those steps make up a geometric series.

- ***p*-series:** The *p*-series $\sum \frac{1}{n^p}$ converges if $p > 1$ and diverges if $p \leq 1$.

- **Telescoping series:** The telescoping series, written as $(h_1 - h_2) + (h_2 - h_3) + (h_3 - h_4) + \ldots + (h_n - h_{n+1})$, converges if h_{n+1} converges. In that case, the series converges to $h_1 - \lim_{n \to \infty} h_{n+1}$. If h_{n+1} diverges, so does the series. This series is very rare, so I won't make you practice any problems.

When analyzing the series in this section and the rest of the chapter, remember that multiplying a series by a constant never affects whether it converges or diverges. For example, if $\sum_{n=1}^{\infty} u_n$ converges, then so will $1000 \cdot \sum_{n=1}^{\infty} u_n$. Disregarding any number of initial terms also has no affect on convergence or divergence: If $\sum_{n=1}^{\infty} u_n$ diverges, so will $\sum_{n=982}^{\infty} u_n$.

Q. Does $1 + \frac{1}{2} + \frac{1}{4} + \frac{1}{8} + \frac{1}{16} + \ldots$ converge or diverge? And if it converges, what does it converge to?

A. Each term is the preceding one multiplied by $\frac{1}{2}$. This is, therefore, a geometric series with $r = \frac{1}{2}$. The first term, a, equals one, so the series converges to $\frac{a}{1-r} = \frac{1}{1 - \frac{1}{2}} = 2$.

Q. Does $\sum \frac{1}{\sqrt{n}}$ converge or diverge?

A. $\sum \frac{1}{\sqrt{n}}$ is the *p*-series $\sum \frac{1}{n^{1/2}}$ where $p = \frac{1}{2}$. Because $p < 1$, the series diverges.

3. Does $.008 - .006 + .0045 - .003375 + .00253125 - \ldots$ converge or diverge? If it converges, what's the infinite sum?

Solve It

4. Does $\frac{1}{2} + \frac{1}{4} + \frac{1}{8} + \frac{1}{12} + \frac{1}{16} + \frac{1}{20} + \ldots$ converge or diverge?

Solve It

5. Does $\sum_{n=1}^{\infty} \frac{1}{n}$ converge or diverge?

Solve It

6. Does $1 + \frac{\sqrt[3]{2}}{2} + \frac{\sqrt[3]{3}}{3} + \frac{\sqrt[3]{4}}{4} + \ldots + \frac{\sqrt[3]{n}}{n}$ converge or diverge?

Solve It

Apples and Oranges . . . and Guavas: Three Comparison Tests

With the three comparison tests, you compare the series in question to a benchmark series. If the benchmark converges, so does the given series; if the benchmark diverges, the given series does as well.

- ✔ **The direct comparison test:** Given that $0 \le a_n \le b_n$ for all n, if $\sum b_n$ converges, so does $\sum a_n$, and if $\sum a_n$ diverges, so does $\sum b_n$.

  This could be called the *well, duhh* test. All it says is that a series with terms equal to or greater than the terms of a divergent series must also diverge, and that a series with terms equal to or less than the terms of a convergent series must also converge.

- ✔ **The limit comparison test:** For two series $\sum a_n$ and $\sum b_n$, if $a_n > 0$, $b_n > 0$ and $\lim_{n \to \infty} \dfrac{a_n}{b_n} = L$, where L is finite and positive, then either both series converge or both diverge.

- ✔ **The integral comparison test:** If $f(x)$ is positive, continuous, and decreasing for all $x \ge 1$ and if $a_n = f(n)$, then $\sum_{n=1}^{\infty} a_n$ and $\int_{1}^{\infty} f(x)\,dx$ either both converge or both diverge. Note that for some strange reason, other books don't refer to this as a comparison test, despite the fact that the logic of the three tests in this section is the same.

Use one or more of the three comparison tests to determine the convergence or divergence of the series in the practice problems. Note that you can often solve these problems in more than one way.

Q. Does $\sum_{n=2}^{\infty} \dfrac{1}{\ln n}$ converge or diverge?

A. **It diverges.**

Note that the nth term test is no help because $\lim_{n \to \infty} \dfrac{1}{\ln n} = 0$. You know from the p-series rule that $\sum_{n=1}^{\infty} \dfrac{1}{n}$ diverges. $\sum_{n=2}^{\infty} \dfrac{1}{n}$, of course, also diverges. The direct comparison test now tells you that $\sum_{n=2}^{\infty} \dfrac{1}{\ln n}$ must diverge as well because each term of $\sum_{n=2}^{\infty} \dfrac{1}{\ln n}$ is greater than the corresponding term of $\sum_{n=2}^{\infty} \dfrac{1}{n}$.

Q. Does $\sum_{n=2}^{\infty} \dfrac{1}{n^2 - n}$ converge or diverge?

A. It converges.

1. **Try the *n*th term test.**

   No good: $\lim\limits_{n \to \infty} \dfrac{1}{n^2 - n} = 0$

2. **Try the direct comparison test.**

   $\sum_{n=2}^{\infty} \dfrac{1}{n^2 - n}$ resembles $\sum_{n=2}^{\infty} \dfrac{1}{n^2}$, which you know converges by the *p*-series rule. But the direct comparison test is no help because each term of $\sum_{n=2}^{\infty} \dfrac{1}{n^2 - n}$ is *greater* than your known convergent series.

3. **Try the limit comparison test with $\sum_{n=2}^{\infty} \dfrac{1}{n^2}$. Piece o' cake.**

   It's best to put your known, benchmark series in the denominator.

   $$\lim_{n \to \infty} \dfrac{\dfrac{1}{n^2 - n}}{\dfrac{1}{n^2}}$$

   $$= \lim_{n \to \infty} \dfrac{n^2}{n^2 - n}$$

   $$= 1 \quad \left(\text{By the horizontal asymptote rule} \right)$$

   Because the limit is finite and positive and because $\sum_{n=2}^{\infty} \dfrac{1}{n^2}$ converges, $\sum_{n=2}^{\infty} \dfrac{1}{n^2 - n}$ also converges.

Q. Does $\sum_{n=2}^{\infty} \dfrac{1}{n \sqrt{\ln n}}$ converge or diverge?

A. It diverges.

Tip: If you can see that you'll be able to integrate the series expression, you're home free. So always ask yourself whether you can use the integral comparison test.

1. **Ask yourself whether you know how to integrate this expression.**

   Sure. It's an easy *u*-substitution.

2. **Do the integration.**

   $$\int_{2}^{\infty} \dfrac{1}{x \sqrt{\ln x}}\, dx$$

   $$= \lim_{c \to \infty} \int_{2}^{c} \dfrac{1}{x \sqrt{\ln x}}\, dx$$

   $u = \ln x \quad$ when $x = 2, \ u = \ln 2$

   $du = \dfrac{1}{x}\, dx \quad$ when $x = c, \ u = \ln c$

   $$= \lim_{c \to \infty} \int_{\ln 2}^{\ln c} u^{-1/2}\, du$$

   $$= \lim_{c \to \infty} \left[2 u^{1/2} \right]_{\ln 2}^{\ln c}$$

   $$= \lim_{c \to \infty} \left(2 \sqrt{\ln c} - 2 \sqrt{\ln 2} \right)$$

   $$= \infty$$

   Because this improper integral diverges, so does the companion series.

7. $\displaystyle\sum_{n=1}^{\infty}\frac{10(.9)^n}{\sqrt{n}}$

Solve It

8. $\displaystyle\sum_{n=1}^{\infty}\frac{1.1^n}{10n}$

Solve It

9. $\dfrac{1}{1001}+\dfrac{1}{2001}+\dfrac{1}{3001}+\dfrac{1}{4001}+\dots$

Solve It

10. $\displaystyle\sum_{n=1}^{\infty}\frac{1}{n+\sqrt{n}+\ln n}$

Solve It

*11. $\sum_{n=2}^{\infty} \frac{1}{n^3 - (\ln n)^3}$

Solve It

*12. $\sum_{n=2}^{\infty} \frac{1}{n \ln n + \sin n}$

Solve It

13. $\sum_{n=1}^{\infty} \frac{n^2}{e^{n^3}}$

Solve It

*14. $\sum_{n=1}^{\infty} \frac{n^3}{n!}$ (Given that $\sum \frac{1}{n!}$ converges.)

Solve It

Ratiocinating the Two "R" Tests

Here you practice the ratio test and the root test. With both tests, a result less than 1 means that the series in question converges; a result greater than 1 means that the series diverges; and a result of 1 tells you nothing.

✔ **The ratio test:** Given a series $\sum u_n$, consider the limit of the ratio of a term to the previous term, $\lim\limits_{n \to \infty} \dfrac{u_{n+1}}{u_n}$. If this limit is less than 1, the series converges. If it's greater than 1 (this includes ∞), the series diverges. And if it equals 1, the ratio test tells you nothing.

✔ **The root test:** Note its similarity to the ratio test. Given a series $\sum u_n$, consider the limit of the nth root of the nth term, $\lim\limits_{n \to \infty} \sqrt[n]{u_n}$. If this limit is less than 1, the series converges. If it's greater (including ∞), the series diverges. And if it equals 1, the root test says nothing.

The ratio test is a good test to try if the series involves factorials like $n!$ or where n is in the power like 2^n. The root test also works well when the series contains nth powers. If you're not sure which test to try first, start with the ratio test — it's often the easier to use.

Sometimes it's useful to have an idea about the convergence or divergence of a series before using one of the tests to prove convergence or divergence.

Q. Does $\sum\limits_{n=1}^{\infty} \dfrac{n}{2^n}$ converge or diverge?

A. Try the ratio test.

$$\lim_{n \to \infty} \frac{\dfrac{n+1}{2^{(n+1)}}}{\dfrac{n}{2^n}} = \lim_{n \to \infty} \frac{2^n(n+1)}{2^{n+1} \cdot n} = \lim_{n \to \infty} \frac{n+1}{2n} = \frac{1}{2}$$

Because this is less than 1, the series converges.

Q. Does $\sum\limits_{n=1}^{\infty} \dfrac{5^{3n+4}}{n^{3n}}$ converge or diverge?

A. Consider the limit of the nth root of the nth term:

$$\lim_{n \to \infty} \sqrt[n]{\frac{5^{3n+4}}{n^{3n}}} = \lim_{n \to \infty} \left(\frac{5^{3n+4}}{n^{3n}} \right)^{1/n} = \lim_{n \to \infty} \frac{5^{3+4/n}}{n^3} = 0$$

Because this limit is less than 1, the series converges.

15. $\displaystyle\sum_{n=1}^{\infty}\frac{1}{\left(\ln\left(n+2\right)\right)^{n}}$

Solve It

16. $\displaystyle\sum_{n=1}^{\infty}\frac{n^{\sqrt{n}}}{\sqrt{n}^{n}}$

Solve It

*17. $\displaystyle\sum_{n=1}^{\infty}\frac{n!}{n^{n}}$

Solve It

*18. $\displaystyle\sum_{n=1}^{\infty}n\left(\frac{3}{4}\right)^{n}$

Solve It

19. $\sum_{n=1}^{\infty} \dfrac{n^{\sqrt{n}}}{n!}$

Solve It

20. $\sum_{n=1}^{\infty} \dfrac{n!}{4^{n}}$

Solve It

He Loves Me, He Loves Me Not:
Alternating Series

Alternating series look just like any other series except that they contain an extra $(-1)^{n}$ or $(-1)^{n+1}$. This extra term causes the terms of the series to *alternate* between positive and negative.

An alternating series converges if two conditions are met:

1. Its nth term converges to zero.

2. Its terms are non-increasing — in other words, each term is either smaller than or the same as its predecessor (ignoring the minus sign).

For the problems in this section, determine whether the series converges or diverges. If it converges, determine whether the convergence is absolute or conditional.

If you take a convergent alternating series and make all the terms positive and it still converges, then the alternating series is said to converge *absolutely*. If, on the other hand, the series of positive terms diverges, then the alternating series converges *conditionally*.

Q. $\displaystyle\sum_{n=1}^{\infty}(-1)^{n}\frac{1}{\sqrt{n}}$

A. **The series is conditionally convergent.**

If you make this a series of positive terms, it becomes a p-series with $p=\frac{1}{2}$, which you know diverges. Thus, the above alternating series is not absolutely convergent. It is, however, conditionally convergent because it obviously satisfies the two conditions of the alternating series test:

1. **The nth term converges to zero.**

$$\lim_{n\to\infty}\frac{1}{\sqrt{n}}=0$$

2. **The terms are non-increasing.**

The series is thus conditionally convergent.

---

21. $\displaystyle\sum_{n=1}^{\infty}(-1)^{n+1}\frac{n+1}{3n+1}$

Solve It

22. $\displaystyle\sum_{n=3}^{\infty}(-1)^{n}\frac{n+1}{n^{2}-2}$

Solve It

Solutions to Infinite Series

1 $\sum_{n=1}^{\infty} \dfrac{2n^2 - 9n - 8}{5n^2 + 20n + 12}$ **diverges.** You know (vaguely remember?) from Chapter 4 on limits that

$\lim\limits_{n \to \infty} \dfrac{2n^2 - 9n - 8}{5n^2 + 20n + 12} = \dfrac{2}{5}$ by the horizontal asymptote rule. Because the limit doesn't converge to zero, the nth term test tells you that the series diverges.

2 $\sum_{n=1}^{\infty} \dfrac{1}{n}$ **converges to zero . . . *NOT.*** It should be obvious that $\lim\limits_{n \to \infty} \dfrac{1}{n} = 0$. If you conclude that the

series, $\sum_{n=1}^{\infty} \dfrac{1}{n}$, must therefore converge by the nth term test, I've got some bad news and some

good news for you. The bad news is that you're wrong — you have to use the *p*-series test to find out whether this converges or not (check out the solution to problem 5). The good news is that you made this mistake here instead of on a test.

Don't forget that the nth term test is no help in determining the convergence or divergence of a series when the underlying sequence converges to zero.

3 $.008 - .006 + .0045 - .003375 + .00253125 - \ldots$ **converges to** $\dfrac{4}{875}$.

1. Determine the ratio of the second term to the first term: $\dfrac{-.006}{.008} = -\dfrac{3}{4}$.

2. Check to see whether all the other ratios of the other pairs of consecutive terms equal $\dfrac{-3}{4}$.

   $\dfrac{.0045}{-.006} = -\dfrac{3}{4}$? check. $\qquad \dfrac{-.003375}{.0045} = -\dfrac{3}{4}$? check. $\qquad \dfrac{.00253125}{-.003375} = -\dfrac{3}{4}$? check.

   Voila! A geometric series with $r = -\dfrac{3}{4}$.

3. Apply the geometric series rule.

   Because $-1 < |r| < 1$, the series converges to

   $$\dfrac{a}{1 - r} = \dfrac{.008}{1 - \left(-\dfrac{3}{4}\right)} = \dfrac{4}{875}$$

"r" is for ratio, but you may prefer, as I do, to think of r ($-\dfrac{3}{4}$ in this problem) as a *multiplier* because it's the number you multiply each term by to obtain the next.

4 $\dfrac{1}{2} + \dfrac{1}{4} + \dfrac{1}{8} + \dfrac{1}{12} + \dfrac{1}{16} + \dfrac{1}{20} + \ldots$?

1. Find the first ratio.

   $\dfrac{\frac{1}{4}}{\frac{1}{2}} = \dfrac{1}{2}$

2. Test the other pairs.

   $\dfrac{\frac{1}{8}}{\frac{1}{4}} = \dfrac{1}{2}$? check. $\qquad \dfrac{\frac{1}{12}}{\frac{1}{8}} = \dfrac{1}{2}$? no.

Thus, this is *not* a geometric series, and the geometric series rule does not apply. Can you guess whether this series converges or not (assuming the pattern 8, 12, 16, 20 continues)? You can prove that this series diverges by using the limit comparison test (see problem 10) with the harmonic series (see problem 5).

5 $\sum_{n=1}^{\infty} \frac{1}{n}$ **diverges.**

With the p-series rule, you can now solve problem 2. $\sum \frac{1}{n}$, called the *harmonic series* $(1 + \frac{1}{2} + \frac{1}{3} + \frac{1}{4} + \frac{1}{5} + \ldots + \frac{1}{n})$, is probably the most important p-series. Because $p = 1$, the p-series rule tells you that the harmonic series diverges.

6 $1 + \frac{\sqrt[4]{2}}{2} + \frac{\sqrt[4]{3}}{3} + \frac{\sqrt[4]{4}}{4} + \ldots + \frac{\sqrt[4]{n}}{n}$ **diverges.**

This may not look like a p-series, but you can't always judge a book by its cover.

1. Rewrite the terms with exponents instead of roots: $= 1 + \frac{2^{1/4}}{2} + \frac{3^{1/4}}{3} + \frac{4^{1/4}}{4} + \ldots + \frac{n^{1/4}}{n}$.

2. Use ordinary laws of exponents to move each numerator to the denominator.

$$= 1 + \frac{1}{2^{3/4}} + \frac{1}{3^{3/4}} + \frac{1}{4^{3/4}} + \ldots + \frac{1}{n^{3/4}}$$

3. Apply the p-series rule. You've got a p-series with $p = \frac{3}{4}$, so this series diverges.

7 $\sum_{n=1}^{\infty} \frac{10(.9)^n}{\sqrt{n}}$ **converges.**

1. Look in the summation expression for a series you recognize that can be used for your benchmark series. You should recognize $\sum .9^n$ as a convergent geometric series, because r, namely 0.9, is between 0 and 1.

2. Use the direct comparison test to compare $\sum_{n=1}^{\infty} \frac{10(.9)^n}{\sqrt{n}}$ to $\sum_{n=1}^{\infty} .9^n$. First, you can pull the 10 out and ignore it because multiplying a series by a constant has no effect on its convergence or divergence, giving you $\sum_{n=1}^{\infty} \frac{.9^n}{\sqrt{n}}$.

Now, because each term of $\sum_{n=1}^{\infty} \frac{.9^n}{\sqrt{n}}$ is less than or equal to the corresponding term of the *convergent* series $\sum_{n=1}^{\infty} .9^n$, $\sum_{n=1}^{\infty} \frac{.9^n}{\sqrt{n}}$ has to converge as well. Finally, because $\sum_{n=1}^{\infty} \frac{.9^n}{\sqrt{n}}$ converges, so does $\sum_{n=1}^{\infty} \frac{10(.9)^n}{\sqrt{n}}$.

8 $\sum_{n=1}^{\infty} \frac{1.1^n}{10n}$ **diverges.**

1. Find an appropriate benchmark series. Like in problem 7, there is a geometric series in the numerator, $\sum_{n=1}^{\infty} 1.1^n$. By the geometric series rule, it diverges. But unlike problem 7, this doesn't help you, because the given series is *less than* this *divergent* geometric series. Use the series in the denominator instead.

$\sum_{n=1}^{\infty} \frac{1.1^n}{10n} = \frac{1}{10} \sum_{n=1}^{\infty} \frac{1.1^n}{n}$. The denominator of $\sum_{n=1}^{\infty} \frac{1.1^n}{n}$ is the divergent p-series $\sum_{n=1}^{\infty} \frac{1}{n}$.

2. Apply the direct comparison test. Because each term of $\sum_{n=1}^{\infty} \frac{1.1^n}{n}$ is *greater* than the corresponding term of the *divergent* series $\sum_{n=1}^{\infty} \frac{1}{n}$, $\sum_{n=1}^{\infty} \frac{1.1^n}{n}$ diverges as well — and therefore so does $\sum_{n=1}^{\infty} \frac{1.1^n}{10n}$.

9 $\frac{1}{1001} + \frac{1}{2001} + \frac{1}{3001} + \frac{1}{4001} + \ldots$ **diverges.**

1. Ask yourself what this series resembles: It's the divergent harmonic series: $\frac{1}{1} + \frac{1}{2} + \frac{1}{3} + \frac{1}{4} + \ldots$.

2. Multiply the given series by 1001 so that you can compare it to the harmonic series.

$$1001\left(\frac{1}{1001} + \frac{1}{2001} + \frac{1}{3001} + \frac{1}{4001} + \ldots\right) = \frac{1001}{1001} + \frac{1001}{2001} + \frac{1001}{3001} + \frac{1001}{4001} + \ldots$$

3. Use the direct comparison test. It's easy to show that the terms of the series in Step 2 are greater than or equal to the terms of the divergent p-series, so it, and thus your given series, diverges as well.

10 $\quad \sum\limits_{n=1}^{\infty} \dfrac{1}{n + \sqrt{n} + \ln n}$ **diverges.**

Try the limit comparison test: Use the divergent harmonic series $\sum\limits_{n=1}^{\infty} \dfrac{1}{n}$, as your benchmark.

$$\lim_{n \to \infty} \dfrac{\dfrac{1}{n + \sqrt{n} + \ln n}}{\dfrac{1}{n}}$$

$$= \lim_{n \to \infty} \dfrac{n}{n + \sqrt{n} + \ln n}$$

$$= \lim_{n \to \infty} \dfrac{1}{1 + \dfrac{1}{2\sqrt{n}} + \dfrac{1}{n}} \quad \text{(By L'Hôpital's Rule)}$$

$$= 1$$

Because the limit is finite and positive, the limit comparison test tells you that $\sum\limits_{n=1}^{\infty} \dfrac{1}{n + \sqrt{n} + \ln n}$

diverges with the benchmark series. By the way, you could do this problem with the direct comparison test as well. Do you see how? **_Hint:_** You can use the harmonic series as your benchmark, but you have to tweak it first.

*11 $\quad \sum\limits_{n=1}^{\infty} \dfrac{1}{n^3 - (\ln n)^3}$ **converges.**

1. Do a quick check to see whether the direct comparison test will give you an immediate answer.

It doesn't because $\sum\limits_{n=1}^{\infty} \dfrac{1}{n^3 - (\ln n)^3}$ is _larger_ than the known _convergent_ p-series $\sum\limits_{n=1}^{\infty} \dfrac{1}{n^3}$.

2. Try the limit comparison test with $\sum\limits_{n=1}^{\infty} \dfrac{1}{n^3}$ as your benchmark.

$$\lim_{n \to \infty} \dfrac{\dfrac{1}{n^3 - (\ln n)^3}}{\dfrac{1}{n^3}}$$

$$= \lim_{n \to \infty} \dfrac{n^3}{n^3 - (\ln n)^3}$$

$$= \lim_{n \to \infty} \dfrac{1}{1 - \dfrac{(\ln n)^3}{n^3}}$$

$$= \lim_{n \to \infty} \dfrac{1}{1 - \left(\dfrac{\ln n}{n}\right)^3}$$

$$= \dfrac{1}{1 - \lim\limits_{n \to \infty}\left(\dfrac{\ln n}{n}\right)^3} \quad \text{(Just take my word for it.)}$$

$$= \dfrac{1}{1 - \left(\lim\limits_{n \to \infty}\dfrac{\ln n}{n}\right)^3} \quad \text{(Just take my word for it.)}$$

$$= \dfrac{1}{1 - \left(\lim\limits_{n \to \infty}\dfrac{\frac{1}{n}}{1}\right)^3} \quad \text{(L'Hôpital's Rule from Chapter 12)}$$

$$= 1$$

Because this is finite and positive, the limit comparison test tells you that $\sum\limits_{n=1}^{\infty} \dfrac{1}{n^3 - (\ln n)^3}$ converges with the benchmark series.

* 12 $\sum\limits_{n=2}^{\infty} \dfrac{1}{n\ln n + \sin n}$ **diverges.**

1. You know you can integrate $\int \dfrac{1}{x\ln x}$ with a simple u-substitution, so do it, and then you'll be able to use the integral comparison test.

$$\int_{2}^{\infty} \dfrac{dx}{x\ln x} \qquad\qquad u = \ln x \quad \text{when } x = 2, \ u = \ln 2$$

$$= \lim_{c\to\infty} \int_{2}^{c} \dfrac{dx}{x\ln x} \qquad du = \dfrac{dx}{x} \quad \text{when } x = c, \ u = \ln c$$

$$= \lim_{c\to\infty} \int_{\ln 2}^{\ln c} \dfrac{du}{u}$$

$$= \lim_{c\to\infty} \left[\ln u\right]_{\ln 2}^{\ln c}$$

$$= \lim_{c\to\infty} \left(\ln\left(\ln c\right) - \ln\left(\ln 2\right)\right)$$

$$= \infty$$

By the integral comparison test, $\sum\limits_{n=2}^{\infty} \dfrac{1}{n\ln n}$ diverges with its companion improper integral, $\int_{2}^{\infty} \dfrac{dx}{x\ln x}$.

2. Try the direct comparison test. Won't work yet because $\dfrac{1}{n\ln n + \sin n}$ is sometimes *less than* $\dfrac{1}{n\ln n}$.

3. Try multiplication by a constant (always easy to do and always a good thing to try).

$\sum\limits_{n=2}^{\infty} \dfrac{1}{n\ln n}$ diverges, thus so does $\dfrac{1}{2}\sum\limits_{n=2}^{\infty} \dfrac{1}{n\ln n} = \sum\limits_{n=2}^{\infty} \dfrac{1}{2n\ln n}$.

4. Now try the direct comparison test again. It's easy to show that $\dfrac{1}{n\ln n + \sin n}$ is always *greater than* $\dfrac{1}{2n\ln n}$ $(n \geq 2)$, and thus the direct comparison test tells you that $\sum\limits_{n=2}^{\infty} \dfrac{1}{n\ln n + \sin n}$ must diverge with $\sum\limits_{n=2}^{\infty} \dfrac{1}{2n\ln n}$.

13 $\sum\limits_{n=1}^{\infty} \dfrac{n^2}{e^{n^3}}$ **converges.**

This is ready-made for the integral test:

$$\int_{1}^{\infty} \dfrac{x^2}{e^{x^3}}\,dx = \lim_{c\to\infty} \int_{1}^{c} \dfrac{x^2}{e^{x^3}}\,dx = \lim_{c\to\infty} \dfrac{1}{3}\int_{1}^{c^3} \dfrac{du}{e^u} = \dfrac{1}{3}\lim_{c\to\infty}\left[-e^{-u}\right]_{1}^{c^3} = -\dfrac{1}{3}\lim_{c\to\infty}\left(\dfrac{1}{e^{c^3}} - \dfrac{1}{e}\right) = \dfrac{1}{3e}$$

Because the integral converges, so does the series.

* 14 $\sum\limits_{n=1}^{\infty} \dfrac{n^3}{n!}$ **converges.**

1. Try the limit comparison test with the convergent series, $\sum\limits_{n=1}^{\infty} \dfrac{1}{n!}$, as the benchmark.

$$\lim_{n\to\infty} \dfrac{\dfrac{n^3}{n!}}{\dfrac{1}{n!}} = \lim_{n\to\infty} \dfrac{n!n^3}{n!} = \infty \text{ No good. This result tells you nothing.}$$

2. Try the following nifty trick. Ignore the first three terms of $\sum_{n=1}^{\infty} \frac{n^3}{n!}$, which doesn't affect the convergence or divergence of the series. The series is now $\frac{4^3}{4!} + \frac{5^3}{5!} + \frac{6^3}{6!} + \ldots$, which can be written as $\sum_{n=1}^{\infty} \frac{(n+3)^3}{(n+3)!}$.

3. Try the limit comparison test again.

$$\lim_{n \to \infty} \frac{\frac{(n+3)^3}{(n+3)!}}{\frac{1}{n!}}$$

$$= \lim_{n \to \infty} \frac{n!(n+3)^3}{(n+3)!}$$

$$= \lim_{n \to \infty} \frac{(n+3)^3}{(n+3)(n+2)(n+1)}$$

$$= \lim_{n \to \infty} \frac{n^3 + \text{lesser powers of } n}{n^3 + \text{lesser powers of } n}$$

$$= 1 \quad (\text{by the horizontal asymptote rule})$$

Thus, $\sum_{n=1}^{\infty} \frac{(n+3)^3}{(n+3)!}$ converges by the limit comparison test. And because $\sum_{n=1}^{\infty} \frac{n^3}{n!}$ is the same series except for its first three terms, it converges as well.

15 $\sum_{n=1}^{\infty} \frac{1}{\left(\ln(n+2)\right)^n}$ **converges.**

$$\lim_{n \to \infty} \left(\frac{1}{\left(\ln(n+2)\right)^n} \right)^{1/n}$$

Try the root test: $= \lim_{n \to \infty} \frac{1}{\ln(n+2)}$

$$= 0$$

This is less than 1, so the series converges.

16 $\sum_{n=1}^{\infty} \frac{n^{\sqrt{n}}}{\sqrt{n}^n}$ **converges.**

Try the root test again:

$$\lim_{n \to \infty} \left(\frac{n^{\sqrt{n}}}{\sqrt{n}^n} \right)^{1/n} = \lim_{n \to \infty} \frac{n^{\sqrt{n}/n}}{n^{1/2}} = \lim_{n \to \infty} \frac{1}{n^{1/2 - \sqrt{n}/n}} = 0$$

Thus the series converges.

*17 $\sum\limits_{n=1}^{\infty} \dfrac{n!}{n^n}$ **converges.**

There's a factorial, so try the ratio test:

$$\lim_{n \to \infty} \frac{\dfrac{(n+1)!}{(n+1)^{n+1}}}{\dfrac{n!}{n^n}}$$

$$= \lim_{n \to \infty} \frac{(n+1)!\, n^n}{n!\,(n+1)^{n+1}}$$

$$= \lim_{n \to \infty} \frac{(n+1)\, n^n}{(n+1)^{n+1}}$$

$$= \lim_{n \to \infty} \frac{n^n}{(n+1)^n}$$

$$= \lim_{n \to \infty} \left(\frac{n}{n+1}\right)^n$$

Finish in the right
column with
logarithmic
differentiation.

$$y = \lim_{n \to \infty} \left(\frac{n}{n+1}\right)^n$$

$$\ln y = \ln\left(\lim_{n \to \infty}\left(\frac{n}{n+1}\right)^n\right)$$

$$= \lim_{n \to \infty}\left(\ln\left(\frac{n}{n+1}\right)^n\right)$$

$$= \lim_{n \to \infty}\left(\frac{\ln\left(\dfrac{n}{n+1}\right)}{\dfrac{1}{n}}\right)$$

$$= \lim_{n \to \infty}\left(\frac{\dfrac{n+1}{n}\cdot\dfrac{n+1-n}{(n+1)^2}}{\dfrac{-1}{n^2}}\right) \quad \text{(L'Hôpital's Rule)}$$

$$= \lim_{n \to \infty}\left(\frac{-n}{n+1}\right)$$

$$\ln y = -1$$

$$y = e^{-1}$$

Because this is less than 1,
the series converges.

*18 $\sum\limits_{n=1}^{\infty} n\left(\dfrac{3}{4}\right)^n$ **converges.**

Rewrite this so it's one big nth power: $\sum\limits_{n=1}^{\infty}\left(n^{1/n}\cdot\dfrac{3}{4}\right)^n$. Now look at the limit of the nth root.

$$\lim_{n \to \infty}\left(\left(\frac{3}{4}\,n^{1/n}\right)^n\right)^{1/n}$$

$$= \lim_{n \to \infty}\frac{3}{4}\,n^{1/n}$$

$$= \frac{3}{4}\lim_{n \to \infty} n^{1/n} \quad \text{(An unacceptable L'Hopital's Rule case: } \infty^0\text{)}$$

$$\ln y = \ln\left(\frac{3}{4}\lim_{n \to \infty} n^{1/n}\right)$$

$$= \ln\frac{3}{4} + \lim_{n \to \infty}\left(\ln n^{1/n}\right)$$

$$= \ln\frac{3}{4} + \lim_{n \to \infty}\frac{\ln n}{n}$$

$$= \ln\frac{3}{4} + \lim_{n \to \infty}\frac{\dfrac{1}{n}}{1} \quad \text{(L'Hopital's Rule)}$$

$$\ln y = \ln\frac{3}{4}$$

$$y = \frac{3}{4}$$

Thus the limit of the nth root is $\dfrac{3}{4}$ and so the series converges.

19 $\sum_{n=1}^{\infty} \dfrac{n^{\sqrt{n}}}{n!}$ **converges.**

Try the ratio test:

$$\lim_{n \to \infty} \dfrac{\dfrac{(n+1)^{\sqrt{n+1}}}{(n+1)!}}{\dfrac{n^{\sqrt{n}}}{n!}} = \lim_{n \to \infty} \dfrac{n!\,(n+1)^{\sqrt{n+1}}}{(n+1)!\,n^{\sqrt{n}}} = \lim_{n \to \infty} \dfrac{(n+1)^{\sqrt{n+1}}}{(n+1)\,n^{\sqrt{n}}} = \lim_{n \to \infty} \dfrac{(n+1)^{\sqrt{n+1}-1}}{n^{\sqrt{n}}} = 0$$

By the ratio test, the series converges.

20 $\sum_{n=1}^{\infty} \dfrac{n!}{4^n}$ **diverges.**

Try the ratio test: $\lim_{n \to \infty} \dfrac{\dfrac{(n+1)!}{4^{n+1}}}{\dfrac{n!}{4^n}} = \lim_{n \to \infty} \dfrac{(n+1)!\,4^n}{n!\,4^{n+1}} = \lim_{n \to \infty} \dfrac{n+1}{4} = \infty$

Thus the series diverges.

21 $\sum_{n=1}^{\infty} (-1)^{n+1} \dfrac{n+1}{3n+1}$ **diverges.** This one is a no-brainer, because $\lim_{n \to \infty} \dfrac{n+1}{3n+1} = \dfrac{1}{3}$, the first condition of the alternating series test is not satisfied, which means that both the alternating series and the series of positive terms are divergent.

*22 $\sum_{n=3}^{\infty} (-1)^n \dfrac{n+1}{n^2-2}$ **converges conditionally.**

Check the two conditions of the alternating series test:

1. $\lim_{n \to \infty} \dfrac{n+1}{n^2-2}$

$= \lim_{n \to \infty} \dfrac{1}{2n}$ (L'Hôpital's Rule)

$= 0$ Check.

2. Are the terms non-increasing?

$\dfrac{n+1}{n^2-2} \geq \dfrac{(n+1)+1}{(n+1)^2-2}$?

$\dfrac{n+1}{n^2-2} \geq \dfrac{n+2}{n^2+2n-1}$?

$(n+1)(n^2+2n-1) \geq (n+2)(n^2-2)$?

$n^3+2n^2-n+n^2+2n-1 \geq n^3-2n+2n^2-4$?

$n^3+3n^2+n-1 \geq n^3+2n^2-2n-4$?

$n^2+3n+5 \geq 0$? Check.

Thus the series is at least conditionally convergent. And it is easy to show that it is only conditionally convergent and not absolutely convergent by the direct comparison test. Each term of $\sum_{n=3}^{\infty} \dfrac{n+1}{n^2-2}$ has a larger numerator and a smaller denominator — and is thus greater than the corresponding term of $\sum_{n=3}^{\infty} \dfrac{n}{n^2}$. $\sum_{n=3}^{\infty} \dfrac{n}{n^2}$ is the same as the divergent harmonic series, $\sum_{n=3}^{\infty} \dfrac{1}{n}$, and therefore $\sum_{n=3}^{\infty} \dfrac{n+1}{n^2-2}$ is also divergent.

Part V
The Part of Tens

The 5th Wave — By Rich Tennant

"Watch this, Ruth. Steady... steady... calculate the volume of the stein first."

In this part . . .

*H*ere I give you ten things you should know about limits and infinite series, ten things you should know about differentiation, and ten things you should know about integration. Remember these 30 things or your name is Mudd.

Chapter 14

Ten Things about Limits, Continuity, and Infinite Series

In This Chapter

▶ When limits, continuity, and derivatives *don't* exist

▶ Ten tests for convergence

In this very short chapter, I give you two great mnemonics for memorizing a great deal about limits, continuity, derivatives, and infinite series. If I do say so myself, you're getting a lot of bang for your buck here.

The 33333 Mnemonic

This mnemonic is a memory aid for limits, continuity, and derivatives. First, note that I've put the word "limil" under the five threes. That's "limit" with the *t* changed to an *l*. Also note the nice parallel between "limil" and the second mnemonic in this chapter, the 13231 mnemonic — in both cases, you've got two pairs surrounding a single letter or number in the center.

3 3 3 3 3

l i m i l

First 3 over the "*l*": 3 parts to the definition of a *l*imit

You can find the formal definition of a limit in Chapter 3. This mnemonic helps you remember that it's got three parts. And — take my word for it — just that is usually enough to help you remember what the three parts are. Try it.

Fifth 3 over the "l": 3 cases where a limit fails to exist

The three cases are

- At a vertical asymptote. This is an *infinite discontinuity*.
- At a *jump discontinuity*.
- With the limit at infinity or negative infinity of an *oscillating function* like $\lim\limits_{x \to \infty} \cos x$ where the function keeps oscillating up and down forever, never honing in on a single *y*-value.

Second 3 over the "i": 3 parts to the definition of continuity

First notice the oh-so-clever fact that the letter *i* can't be drawn without taking your pen off the paper and thus that it's not *continuous*. This will help you remember that the second and fourth 3s concern continuity.

The three-part, formal definition of continuity is in Chapter 3. The mnemonic will help you remember that it's got three parts. And — just like with the definition of a limit — that's enough to help you remember what the three parts are.

Fourth 3 over the "i": 3 cases where continuity fails to exist

The three cases are

- A *removable discontinuity* — the highfalutin calculus term for a *hole*.
- An *infinite discontinuity*.
- A *jump discontinuity*.

Third 3 over the "m": 3 cases where a derivative fails to exist

Note that *m* often stands for *slope*, right? And the slope is the same thing as the derivative. The three cases where it fails are

- At any type of *discontinuity*.
- At a *cusp*: a sharp point or corner along a function (this only occurs in weird functions).
- At a *vertical tangent*. (A vertical line has an undefined slope and thus an undefined derivative.)

The 13231 Mnemonic

This mnemonic helps you remember the ten tests for the convergence or divergence of an infinite series covered in Chapter 13. $1 + 3 + 2 + 3 + 1 = 10$. Got it?

First 1: The nth term test of divergence

For any series, if the nth term doesn't converge to zero, the series diverges.

Second 1: The nth term test of convergence for alternating series

The real name of this test is the *alternating series test*. But I'm referring to it as the nth term test of convergence because that's a pretty good way to think about it, because it has a lot in common with the nth term test of divergence, because these two tests make nice bookends for the other eight tests, and, last but not least, because it's my book.

An alternating series will converge if 1) its nth term converges to 0, and 2) each term is less than or equal to the preceding term (ignoring the negative signs).

Note the following very nice parallel between the two nth term tests: with the nth term test of divergence, if the nth term fails to converge to zero, then the series fails to converge, but it is *not* true that if the nth term succeeds in converging to zero, then the series must succeed in converging.

With the alternating series nth term test, it's the other way around (sort of). If the test succeeds, then the series succeeds in converging, but it is *not* true that if the test fails, then the series must fail to converge.

First 3: The three tests with names

This "3" helps you remember the three types of series that have names: geometric series (which converge if $|r| < 1$), p-series (which converge if $p > 1$), and telescoping series.

Second 3: The three comparison tests

The *direct* comparison test, the *limit* comparison test, and the *integral* comparison test all work the same way. You compare a given series to a known benchmark series. If the benchmark converges, so does the given series, and ditto for divergence.

The 2 in the middle: The two R tests

The *ratio* test and the *root* test make a coherent pair because for both tests, if the limit is less than 1, the series converges; if the limit is greater than 1, the series diverges; and if the limit equals 1, the test tells you nothing.

Chapter 15

Ten Things You Better Remember about Differentiation

In This Chapter

▶ *Psst*, over here
▶ The difference quotient
▶ Extrema, concavity, and inflection points
▶ The product and quotient rules

In this chapter, I give you ten important things you should know about differentiation. Refer to these pages often. When you get these ten things down cold, you'll have taken a not-insignificant step toward becoming a differentiation expert.

The Difference Quotient

The formal definition of a derivative is based on the *difference quotient*:

$f'(x) = \lim\limits_{h \to o} \dfrac{f(x+h) - f(x)}{h}$; this says basically the same thing as $slope = \dfrac{rise}{run}$.

The First Derivative Is a Rate

A first derivative tells you how much y changes per unit change in x. For example, if y is in *miles* and x is in *hours*, and if at some point along the function, y goes up 3 when x goes over 1, you've got 3 *mph*. That's the rate and that's the derivative.

The First Derivative Is a Slope

In the previous example, when y goes up 3 (the *rise*) as x goes over 1 (the *run*), the slope (*rise*/*run*) at that point of the function would be 3 of course. That's the slope and that's the derivative.

Extrema, Sign Changes, and the First Derivative

When the sign of the first derivative changes from positive to negative or vice-versa, that means that you went up then down (and thus passed over the top of a hill, a *local max*), or you went down then up (and thus passed through the bottom of a valley, a *local min*). In both of these cases of *local extrema*, the first derivative usually will equal zero, though it may be undefined (if the *local extremum* is at a *cusp*). Also, note that if the first derivative equals zero, you may have a *horizontal inflection point* rather than a local extremum.

The Second Derivative and Concavity

A *positive* second derivative tells you that a function is *concave up* (like a spoon holding water or like a smile). A *negative* second derivative means *concave down* (like a spoon spilling water or like a frown).

Inflection Points and Sign Changes in the Second Derivative

Note the very nice parallels between second derivative sign changes and first derivative sign changes described in the section above.

When the sign of the second derivative changes from positive to negative or vice-versa, that means that the concavity of the function changed from up to down or down to up. In either case, you're likely at an *inflection point* (though you could be at a *cusp*). At an inflection point, the second derivative will usually equal zero, though it may be undefined if there's a *vertical tangent* at the inflection point. Also, if the second derivative equals zero, that does not guarantee that you're at an inflection point. The second derivative can equal zero at a point where the function is concave up or down (like, for example, at $x = 0$ on the curve $y = x^4$).

The Product Rule

The derivative of a product of two functions equals the derivative of the first times the second plus the first times the derivative of the second. In symbols, $\frac{d}{dx}(uv) = u'v + uv'$.

The Quotient Rule

The derivative of a quotient of two functions equals the derivative of the top times the bottom *minus* the top times the derivative of the bottom, all over the bottom squared. In symbols, $\frac{d}{dx}\left(\frac{u}{v}\right) = \frac{u'v - uv'}{v^2}$.

Note that the numerator of the quotient rule is identical to the product rule except for the subtraction. For both rules, you begin by taking the derivative of the first thing you read: the *left* function in a product and the *top* function in a quotient.

Linear Approximation

Here's the fancy calculus formula for a linear approximation: $l(x) = f(x_1) + f'(x_1)(x - x_1)$. If trying to memorize this leaves you feeling frustrated, flabbergasted, feebleminded, or flummoxed, or fit to be tied, consider this: It's just an equation of a line, and its meaning is identical to the point-slope form for the equation of a line you learned in algebra I (tweaked a bit): $y = y_1 + m(x - x_1)$.

"PSST," Here's a Good Way to Remember the Derivatives of Trig Functions

Take the last three letters in *PSST* and write down the trig functions that begin with those letters: secant, secant, tangent. Below these write their co-functions, cosecant, cosecant, cotangent (add a negative sign). Then add arrows. The arrows point to the derivatives, for example, the arrow after secant points to its derivative, sec · tan; and the arrow next to tangent points backwards to its derivative, sec². Here you go:

sec $\rightarrow$ sec $\leftarrow$ tan

csc $\rightarrow$ –csc $\leftarrow$ cot

Chapter 16

Ten Things to Remember about Integration If You Know What's Good for You

In This Chapter

▶ Three approximation rules

▶ The Fundamental Theorem of Calculus

▶ Definite and indefinite integrals and antiderivatives

1n this chapter, I give you ten things you should know about integration. If you want to become a fully integrated person (as opposed to a derivative one), integrate these integration rules and make them an integral part of your being.

The Trapezoid Rule

The trapezoid rule will give you a fairly good approximation of the area under a curve in the event that you're unable to — or you choose not to — obtain the exact area with integration.

$$T_n = \frac{b-a}{2n}\left[f(x_0) + 2f(x_1) + 2f(x_2) + 2f(x_3) + \dots + 2f(x_{n-1}) + f(x_n)\right]$$

The Midpoint Rule

An even better area approximation is given by the midpoint rule — it uses rectangles.

$$M_n = \frac{b-a}{n}\left[f\left(\frac{x_0+x_1}{2}\right) + f\left(\frac{x_1+x_2}{2}\right) + f\left(\frac{x_2+x_3}{2}\right) + \dots\dots + f\left(\frac{x_{n-1}+x_n}{2}\right)\right]$$

Simpson's Rule

The best area estimate is given by Simpson's Rule — it uses trapezoid-like shapes that have parabolic tops.

$$S_n = \frac{b-a}{3n}\left[f(x_0) + 4f(x_1) + 2f(x_2) + 4f(x_3) + 2f(x_4) + \dots + 4f(x_{n-1}) + f(x_n)\right]$$

If you already have, say, the midpoint approximation for ten rectangles and the trapezoid approximation for ten trapezoids, you can effortlessly compute the Simpson's Rule approximation for ten curvy-topped "trapezoids" with the following shortcut: $S_{2n} = \dfrac{M_n + M_n + T_n}{3}$. This gives you an extraordinarily good approximation.

The Indefinite Integral

The indefinite integral, $\int f(x)\,dx$, is the family of all antiderivatives of $f(x)$. That's why your answer has to end with "+ C." For example, $\int 2x\,dx$ is the family of all parabolas of the form $x^2 + C$ like $x^2 - 1$, $x^2 + 3$, $x^2 + 10$, and so on. All these are vertical translations of $y = x^2$.

The Fundamental Theorem of Calculus, Take 1

Given an area function A_f that sweeps out area under $f(t)$,

$$A_f(x) = \int_s^x f(t)\,dt,$$

the rate at which area is being swept out is equal to the height of the original function. So, because the rate is the derivative, the derivative of the area function equals the original function:

$$\frac{d}{dx} A_f(x) = f(x).$$

The Fundamental Theorem of Calculus, Take 2

Let F be any antiderivative of the function f; then

$$\int_a^b f(x)\,dx = F(b) - F(b).$$

The Definite Integral

In essence, what all definite integrals, $\int_a^b f(x)\,dx$, do is to add up an infinite number of infinitesimally small pieces of something to get the total amount of the thing between a and b. The expression after the integral symbol, $f(x)\,dx$ (the *integrand*), is always a mathematical expression of a representative piece of the stuff you're adding up.

A Rectangle's Height Equals Top Minus Bottom

If you're adding up rectangles with a definite integral to get the total area between two curves, you need an expression for the height of a representative rectangle. This should be a no-brainer: it's just the rectangle's top y-coordinate minus its bottom y-coordinate.

Area Below the x-Axis Is Negative

If you want, say, the area *below* the x-axis and above $y = \sin x$ between $-\pi$ and 0, the top of a representative rectangle is on the x-axis (the function $y = 0$) and its bottom is on $\sin x$. Thus, the height of the rectangle is $0 - \sin x$, and you use the following definite integral to get the area: $\int_{-\pi}^{0} (0 - \sin x)\, dx$, which equals, of course, $-\int_{-\pi}^{0} \sin x\, dx$. So this *negative* integral gives you the ordinary *positive* area. And that's why an ordinary *positive* integral gives you a *negative* area for the parts of a curve that are below the x-axis.

Integrate in Chunks

When you want the total area between two curves and the "top" function changes because the curves cross each other, you have to use more than one definite integral. Each place the curves cross defines the edge of an area you must integrate separately. (If a function crosses the x-axis, you have to consider $y = 0$ as the second function and the x-intercepts as the crossing points.)

Index

Numbers

13231 mnemonic, 267
33333 mnemonic, 265–266

• A •

absolute convergence, 253–254
absolute extrema, 98–101
acceleration, 131–134
algebra
  review
    problems, 9–11
    solutions, 15–18
  solving limit problems, 39–43
alternating series, 253–254
antiderivatives. *See also* integration
  area function, 177–179
  definition, 179
  finding, 183–187
  Fundamental Theorem of Calculus, 179–182, 274
  guess and check method, 183–184
  negative area, 177–179, 275
  problems, 178–187
  solutions, 188–191
  substitution method, 185–187
approximating
  area under curves, 159–161, 168–170
  linear approximation, 138–139, 271
arc length, 227–228
area, calculating. *See* integration
area function, 177–179
average velocity, 132–134

• C •

calculator, solving limit problems, 44–45
Calculus For Dummies, 61
canceling method, 40–43
chain rule, 75–77
change, calculating speed of. *See* derivatives
chunking integration areas, 275
comparison tests, 247–250
concavity
  curves, 102–105
  second derivative test, 270
conditional convergence, 253–254
conjugate multiplication method, 40–43
constants, derivatives of, 69
continuity
  13231 mnemonic, 267
  33333 mnemonic, 265–266
  definition, 31–32
  problems, 33–36
  solutions, 37–38
convergence/divergence, testing for, 245–254
curves. *See also* derivatives; difference quotient
  approximating area under, 159–161, 168–170
  definite integral, 166–167
  exact areas, 166
  graphs of
    identifying as a function, 19
    review, 20–21
    vertical line test, 19
  index of summation, 162
  integrands, 166
  irregular shapes, 159–161
  long sums, shorthand for, 162–166
  problems, 160–170
  rectangles, 159–161
  Reimann sums, 162–166
  shape analysis
    absolute extrema, 98–101
    concavity, 102–105
    first derivative test, 91–94, 269
    highest/lowest points, 98–101
    hills and valleys, 91–97
    inflections points, 102–105
    local extrema, 91–97
    Mean Value Theorem, 105–107
    problems, 92–107
    rate, calculating average, 105–107
    second derivative test, 95–97, 102
    slope, calculating average, 105–107
    smiles and frowns, 102–105
    solutions, 108–121
  sigma notation, 162–166
  Simpson's rule, 168–170
  solutions, 171–176
  Trapezoid rule, 168–170
cylindrical shell method, 222–226

• D •

definite integral, 166–167, 274
degree/radian conversions, 23
derivative-of-a-constant rule, 69
derivatives. *See also* curves; functions; rates; slope
  absolute extrema, 98–101
  compositions of functions, 75–77
  concavity of curves, 102–105
  of constants, 69
  definition, 59
  of derivatives, 80–81
  difference quotient
    definition, 61, 269
    problems, 62–63
    solutions, 64–67
  first derivative test
    curves, shape analysis, 269
    local extrema, 91–94, 270
    rates, 269
    sign changes, 270
    slopes, 269
  highest/lowest points in curves, 98–101
  hills and valleys in curves, 91–97
  inflections points in curves, 102–105
  local extrema, 91–97
  Mean Value Theorem, 105–107
  problems, 59–61, 92–107
  product of two functions, 72–74
  quotient of two functions, 72–74

rate, calculating average, 105–107

rules for
  chain, 75–77. *See also* implicit differentiation
  derivative of a constant, 69
  high order derivatives, 80–81
  implicit differentiation, 78–79. *See also* chain rule
  power, 69
  problems, 70–81
  product, 72–74
  quotient, 72–74, 270–271
  solutions, 82–89

second derivative test
  concavity, 102, 270
  inflection points, 102, 270
  local extrema, 95–97

slope, calculating average, 105–107

smiles and frowns, curve shape, 102–105

solutions, 64–67, 108–121

symbol for, 7

unable to solve for *y*, 78–79

of variables raised to a power, 69

difference quotient
  definition, 61, 269
  problems, 62–63
  solutions, 64–67

differentiation. *See also* antiderivatives
  acceleration, 131–134
  average velocity, 132–134
  displacement, 131
  distance, 131–134
  linear approximation, 138–139
  lines, 134–137
  negative displacement, 131–134
  negative velocity, 131–134
  normals, 134–137
  optimization, 124–126
  position, 131–134
  problems, 124–139
  related rates, 127–130
  rules for
    chain, 75–77. *See also* implicit differentiation
    derivative of a constant, 69
    high order derivatives, 80–81
    implicit differentiation, 78–79. *See also* chain rule
    power, 69

problems, 70–81
product, 72–74
quotient, 72–74
solutions, 82–89
solutions, 140–156
speed, 132–134
speed and distance traveled, 131–134
tangents, 134–137
velocity, 131–134
direct comparison test, 247–250
disk/washer method, 222–226
displacement, 131
distance, 131–134
divergence/convergence, testing for, 245–254

● E ●

estimation
  area under curves, 159–161, 168–170
  linear approximation, 138–139, 271
exact areas, 166
expressions with trigonometric functions, 196–198

● F ●

factoring method, 40–43
first derivative test
  curves, shape analysis, 269
  local extrema, 91–94, 270
  rates, 269
  sign changes, 270
  slopes, 269
FOILing method, 40–43
fractions
  difference quotient, 7
  partial, 201–204
  review
    problems, 7–8
    solutions, 15–18
frowns and smiles, curve shapes, 102–105
functions. *See also* derivatives; graphs
  average value of, 219–220
  compositions of, 75–77
  continuity
    definition, 31–32
    problems, 33–36
    solutions, 37–38

definition, 19
identifying, 19
length along, 227–228
limits
  definition, 31–32
  problems, 33–36
  solutions, 37–38
product of two, 72–74
quotient of two, 72–74
review
  problems, 20–21
  solutions, 25–28
selecting, 193–196
vertical line test, 19
Fundamental Theorem of Calculus, 179–182, 274

● G ●

geometric series, convergence/divergence testing, 245–246
geometry review
  problems, 12–14
  solutions, 15–18
graphs. *See also* functions
  continuity
    definition, 31–32
    problems, 33–36
    solutions, 37–38
  of curves
    identifying as a function, 19
    review, 20–21
    vertical line test, 19
  limits
    definition, 31–32
    problems, 33–36
    solutions, 37–38
guess and check method, 183–184

● H ●

highest/lowest points, curves, 98–101
high-order-derivatives rule, 80–81
hills and valleys, curves, 91–97
horizontal asymptote, 47
hypotenuse, length calculation, 12

● I ●

implicit differentiation rule, 78–79
improper integrals, 231–233
indefinite integrals, 179–182, 274
index of summation, 162
infinite limits of integration, 231–233
infinite series
13231 mnemonic, 267
33333 mnemonic, 265–266
absolute convergence, 253–254
alternating series, 253–254
comparison tests, 247–250
conditional convergence, 253–254
definition, 243
direct comparison test, 247–250
geometric series, 245–246
integral comparison test, 247–250
limit comparison test, 247–250
problems, 244–254
p-series, 245–246
ratio test, 251–254
root test, 251–254
sequences, 243
solutions, 255–261
telescoping series, 245–246
testing for divergence/convergence, 245–254
inflection points
curves, 102–105
second derivative test, 270
integral comparison test, 247–250
integrals
arc length, 227–228
area calculation, 227–228
areas between curves, 220–221, 274
average value of functions, 219–220
cylindrical shell method, 222–226
definite, 166–167, 274
disk/washer method, 222–226
improper, 231–233
indefinite, 179–182, 274
infinite limit of, 231–233
irregular solids, 222–226
length along a function, 227–228
meat slicer method, 222–226
problems, 220–233

solutions, 234–242
surface of revolution, 227–228
volume calculation, 222–226
integrands, 166
integration
arc length, 227–228
areas. *See also* antiderivatives
approximating, 159–161, 168–170
below the x-axis, 275
chunking, 275
under a curve, 159–161, 168–170, 274–275
definite integral, 166–167
exact areas, 166
index of summation, 162
infinite limits of, 231–233
integrands, 166
irregular shapes, 159–161
length along a function, 227–228
long sums, shorthand for, 162–166
midpoint rule, 273
problems, 160–170, 220–233
rectangles, 159–161
Reimann sums, 162–166
sigma notation, 162–166
Simpson's rule, 168–170, 273–274
solutions, 171–176, 234–242
surface of revolution, 227–228
Trapezoid rule, 168–170, 273
triangles, 12
between curves, 220–221, 274
cylindrical shell method, 222–226
disk/washer method, 222–226
functions, average value of, 219–220
irregular solids, 222–226
meat slicer method, 222–226
by parts. *See also* product rule
definition, 193
expressions with trigonometric functions, 196–198
LIATE mnemonic, 193
partial fractions, 201–204
problems, 194–204
Pythagorean Theorem, 198–201
selecting a function, 193–196

SohCahToa right triangle, 198–201
solutions, 205–216
trigonometric substitution, 198–201
problems, 220–233
solutions, 234–242
volumes, 222–226

● **K** ●

Kasube, Herbert, 193

● **L** ●

least-common-denominator method, 40–43
L'Hôpital's rule, 229–231
LIATE mnemonic, 193
limit comparison test, 247–250
limit problems
horizontal asymptote, 47
limits at infinity, 47–49
rational functions, 47
solutions, 50–56
solving with
algebra, 39–43
calculator, 44–45
canceling method, 40–43
conjugate multiplication method, 40–43
factoring method, 40–43
FOILing method, 40–43
least common denominator method, 40–43
L'Hôpital's rule, 229–231
sandwich method, 46–47
simplification method, 40–43
squeeze method, 46–47
types of expressions, 48–49
limits
13231 mnemonic, 267
33333 mnemonic, 265–266
definition, 31–32
problems, 33–36
solutions, 37–38
linear approximation, 138–139, 271
lines, 134–137
local extrema, 91–97, 270
long sums, shorthand for,

162–166
lowest/highest points, curves,
    98–101

• M •

Mean Value Theorem, 106–107
meat slicer method, 222–226
midpoint rule, 273
mnemonics
    13231, 267
    33333, 265–266
    continuity, 265–267
    derivatives of trigonometry
        functions, 271
    infinite series, 265–267
    limits, 265–267
    PSST, 271

• N •

negative area, 177–179, 275
negative displacement, 131–134
negative velocity, 131–134
normals, 134–137

• O •

13231 mnemonic, 267
optimization, 124–126

• P •

parallelogram, area calculation,
    13
partial fractions, 201–204
position, 131–134
power rule, 69
product rule, 72–74, 270. *See also*
    integration, by parts
p-series, convergence/divergence
    testing, 245–246
PSST mnemonic, 271
Pythagorean Theorem, 198–201

• Q •

quotient rule, 72–74, 270–271

• R •

radian/degree conversions, 23
rates. *See also* derivatives
    calculating average, 105–107
    first derivative test, 269
    related, 127–130
ratio test, 251–254
rational functions, 47
rectangles, area under a curve,
    159–161, 274–275
Reimann sums, 162–166
related rates, 127–130
reverse differentiation. *See* anti-
    derivatives
right triangles, 198–201
root test, 251–254
rules
    for derivatives
        chain, 75–77. *See also* implicit
            differentiation
        derivative of a constant, 69
        high order derivatives, 80–81
        implicit differentiation, 78–79.
            See also chain rule
        power, 69
        problems, 70–81
        product, 72–74
        quotient, 72–74, 270–271
        solutions, 82–89
    for differentiation
        chain, 75–77. *See also* implicit
            differentiation
        derivative of a constant, 69
        high order derivatives, 80–81
        implicit differentiation, 78–79.
            See also chain rule
        power, 69
        problems, 70–81
        product, 72–74
        quotient, 72–74
        solutions, 82–89
    L'Hôpital's, 229–231
    Simpson's, 168–170, 273–274
    Trapezoid, 168–170, 273

• S •

sandwich method, 46–47
second derivative test
    concavity, 102, 270
    inflection points, 102, 270
    local extrema, 95–97

sequences, definition, 243
sigma notation, 162–166
sign changes, first derivative test,
    270
simplification method, 40–43
Simpson's rule, 168–170, 273–274
slope. *See also* derivatives; differ-
    ence quotient
    calculating, 13–14
    calculating average, 105–107
    first derivative test, 269
smiles and frowns, curve shapes,
    102–105
SohCahToa right triangle,
    198–201
speed, 132–134
speed of change, calculating. *See*
    derivatives
squeeze method, 46–47
substitution method, 185–187
surface of revolution, 227–228

• T •

tangents, 134–137
telescoping series, conver-
    gence/divergence testing,
    245–246
33333 mnemonic, 265–266
Trapezoid rule, 168–170, 273
triangles
    area, 12
    geometry review, 12–14
    hypotenuse, 12
    Pythagorean Theorem, 198–201
    right, 198–201
trigonometric functions, in
    expressions, 196–198
trigonometric substitution,
    198–201
trigonometry review
    problems, 22–24
    solutions, 25–28

• V •

valleys and hills, curves, 91–97
velocity, 131–134
vertical tangents, 270
volume, calculating. *See* integra-
    tion